Naturalists' Handbooks 25

Thrips

WILLIAM D.J. KIRK
Department of Biological Sciences,
Keele University

*With colour plates by Anthony J. Hopkins
and illustrations by Jennifer M. Palmer and William D.J. Kirk*

Published for the Company of Biologists Ltd by

The Richmond Publishing Co. Ltd

P.O. Box 963, Slough, SL2 3RS, England

Series editors
S.A. Corbet and R.H.L. Disney

Published by The Richmond Publishing Co. Ltd,
P.O. Box 963, Slough, SL2 3RS
Telephone: 01753 643104

ISBN 0 85546 307 4 Paper
ISBN 0 85546 308 2 Hardcovers

Printed in Great Britain

Contents

Editors' preface

Thrips are tiny insects. They are noticed when they descend from the air in multitudes on summer days or squeeze themselves in behind the glass in picture frames. They attract less attention, even when abundant, in their customary habitats in flowers, on leaves or on fungus-infected bark or leaf litter. The little that we already know about their natural history and behaviour reveals features of special interest to insect biologists and behavioural ecologists. We need to know much more, particularly about the distribution and ecological and evolutionary significance of their dramatic fighting behaviour.

Thrips remain a mysterious group because up to now they have been studied by only a few specialists. A factor limiting their appeal has been the problem of identification. Some species are genuinely difficult to separate, but others can be named fairly easily and some can even be recognised alive in the field. The keys in this book have been designed to enable anyone with access to a good microscope to overcome the taxonomic hurdle in order to explore the unusual natural history and behaviour of thrips.

Thrips are abundant, accessible and enigmatic. We hope this book will enable more people to appreciate and investigate these intriguing animals.

SAC
RHLD
February 1996

Acknowledgements

I am grateful to Sally Corbet, Henry Disney and Maria Kirk for their comments on the text. Richard zur Strassen gave time and advice freely and told me of his observation of fighting in *Aeolothrips intermedius*. The authorities of the Senckenberg Museum, Frankfurt am Main, allowed me to use the thrips collection and borrow specimens. Photographs, drawings and specimens were willingly loaned or given for the illustrations by B.J. Crespi, A.W. Ferguson, T. Lewis, J.L. Osborne, I.L. Terry and L.J. Wadhams. Photographs of thrips and flowers were specially taken for some of the illustrations by Gerald Burgess.

WDJK

1 Introduction

Thrips are very small insects. They usually go unnoticed, even though they are phenomenally abundant all over the world. Smallness has protected thrips from attention, so not much is known about them and few people have even heard of them.

However, they are remarkable animals. They deserve to be noticed rather than overlooked. This book describes the behaviour of a range of common thrips species, including some that are beneficial and some that are harmful. Thrips behaviour is interesting in itself, but an understanding of thrips behaviour can also be useful because it can be exploited to improve the management of thrips pests. An identification key and details of various experimental techniques are provided for those who wish to look at thrips for themselves.

Many people have come across thrips, probably without knowing what they are. Thrips that live on wheat and other cereals are known as thunderbugs or thunderflies. In wheat-growing areas, such as East Anglia, they migrate and land in vast numbers. People can then find themselves covered in tiny black insects that cause a slight itching on the skin. Keen cyclists have probably swallowed quite a few and been momentarily blinded when they have got one in their eye. Thunderbugs are also notorious for insinuating themselves behind the glass of picture frames. Other thrips are pests of peas and onions in the vegetable garden or of cucumbers in glasshouses.

Fig. 1. An adult thripid, as mounted on a microscope slide.

Thrips belong to the order of insects known as Thysanoptera, meaning fringed wings. The name "thrips" is unusual in that it is the singular as well as the plural. There is no such thing as a "thrip"! About 5,000 species of thrips are known throughout the world, and about 160 of these occur in Britain. Most species are only 1–3 mm long when fully grown, but a few "giant" species exist. The largest species in Britain reaches a length of 7 mm, while the largest species in the world reaches 14 mm.

Fig. 2. An adult aeolothripid, as mounted on a microscope slide.

The thrips are divided into two sub-orders, which can be distinguished with the naked eye. Thrips in the sub-order Terebrantia have a blunt or angled end to the body (fig. 1, fig. 2), whereas those in the Tubulifera have a tube at the end of the body (fig. 3).

The sub-order Terebrantia contains seven families (table 1). The biggest of these, in terms of number of species, is the Thripidae (fig. 1), which includes most of the plant-feeding thrips and thus most of the crop pests. The next biggest terebrantian family is the Aeolothripidae (fig. 2), in which there are many predatory species. Adult aeolothripids are easily recognised under the microscope by their broad wings. Some common British species have distinctive black and white striped wings, which can be seen with the naked eye. The other five families contain very few species and are rarely seen.

Fig. 3. An adult phlaeothripid, as mounted on a microscope slide.

Table 1. *The classification of thrips (Mound & Teulon, 1995) and the approximate number of species known in the world and found in Britain.*

Order	Sub-order	Family	World	Britain
Thysanoptera	Terebrantia	Uzelothripidae	1	0
		Merothripidae	15	0
		Aeolothripidae	250	13
		Adiheterothripidae	5	0
		Fauriellidae	4	0
		Heterothripidae	70	0
		Thripidae	1700	107
	Tubulifera	Phlaeothripidae	3000	39

The sub-order Tubulifera contains only one family of thrips, the Phlaeothripidae (fig. 3). This is the largest family in the world. However, it is better represented in tropical regions than in temperate ones, so there are relatively few phlaeothripids in Britain. The species have a wide range of habits. Just over half of the phlaeothripids live on leaf litter or dead wood and feed on fungi, while most of the rest live on green leaves. A few live in flowers, form leaf galls or are predatory.

Only the Thripidae, Aeolothripidae and Phlaeothripidae occur in Britain.

The stages in the life cycle of a thripid are shown on Plates 3 and 4. Members of the Terebrantia insert their eggs into plant tissue, whereas members of the Tubulifera lay their eggs on the surface. The eggs hatch into larvae. The larvae are similar in shape to adults without wings, and their habits are very similar. They are usually whitish yellow, but some are orange or even red. There are two larval stages (larva I and larva II), with a moult in between. The larvae feed and grow rapidly. The larval stages are followed by an intermediate stage (the propupa, sometimes also called the prepupa), and then one pupal stage in Terebrantia (pupa) or two pupal stages in Tubulifera (pupa I and pupa II). Propupae and pupae are generally inactive. They can sometimes move around, but they cannot feed. The propupal and pupal stages are usually spent in the same place as the larval stages. However, this is not possible for thrips that live in flowers because the flowers do not last long enough. Many flower thrips spend the propupal and pupal stages in or on the soil below their host-plant. The adults are usually black, brown or yellow.

Insects such as locusts and aphids develop their immature stages gradually; there is no dramatic change of shape when the adult appears. Their immature stages are usually called nymphs. Insects such as butterflies and moths exhibit a complete metamorphosis; the adults are very different from the young stages. Their immature stages are usually called larvae and pupae. Thrips appear to have a gradual development, but the external appearance is deceptive. There is a major breakdown and reformation of tissues during the pupal stages; the internal development is not gradual. Immature thrips are therefore known as larvae, propupae and pupae, rather than nymphs.

The mouthparts of larval and adult thrips form a prominent cone-shaped structure, the mouthcone, underneath the head (fig. 4). A thrips can stick out a mandible that resembles a microscopic needle and a pair of maxillary stylets that fit together to form a feeding tube. The usual method of feeding is to first pierce a hole with the mandible and then insert the maxillary stylets and suck up liquid food through them. Thrips' mouthparts are often described as "piercing-sucking mouthparts".

Insect mouthparts have evolved from paired structures. Although there is a pair of maxillary stylets, there is only a single mandible. How has the mandible evolved? It looks symmetrical when it is projected, but inside the mouthcone it curves to the left side of the head. The single mandible has evolved from the left mandible, while the right mandible has been lost. All thrips are "left-handed" in this way.

Fig. 4. Head and mouthcone viewed from the side. The part of the front leg that would obstruct the view has been omitted.

On the end of each leg of larvae and adults is an unusual structure called an arolium, a small bladder that can be inflated with blood (fig. 5). The arolium appears to improve adhesion with surfaces, allowing thrips to hold on fast and walk vertically or upside down.

An adult thrips has two pairs of narrow strap-like fringed wings. The hindwings are usually hidden beneath the forewings. Just before flying, thripids and aeolothripids fan out their fringe hairs by arching the abdomen up between the wings (Ellington, 1980*). This behaviour, which can be seen with the naked eye, makes a thrips look like a small scorpion about to sting. The fringe hairs have a two-position notched socket that keeps the hairs either held in or spread out. When spread out, the hairs increase the effective wing area. Adult thrips jump when they start flying. On landing, thrips use their legs to push the hairs back into the resting position. Collapsible wings make it easier for thrips to arrange their wings side by side along the abdomen when not flying. Phlaeothripids have their fringe hairs permanently spread out, but they overlap their wings when at rest, so there is more room for projecting hairs.

arolium

Fig. 5. Tip of fore-leg with expanded arolium on the end.

Thrips have an unusual mode of reproduction. In most insects, as in humans, male and female individuals have two sets of chromosomes, one from their mother and one from their father. This is the diploid condition. The gametes, egg cells and sperms, have only one set of chromosomes. They are haploid. When two gametes fuse at fertilisation, the diploid condition is restored. Eggs have to be fertilised before they will develop and they can become males or females. In most thrips, however, fertilised eggs produce only females whereas unfertilised eggs produce

* References cited under authors' names in the text appear in full in References and further reading on p. 64.

males. Females develop from eggs with two sets of chromosomes (diploid eggs), whereas males develop from eggs with only one set of chromosomes (haploid eggs). This system is also found among ants, bees and wasps and is known as haplodiploidy. It can have interesting consequences (see chapter 4). In populations of some species of thrips, males are absent; the females lay unfertilised eggs which produce only females.

Thrips have evolved a wide range of life styles. Primitive thrips probably lived in leaf litter and fed on fungi. From there they moved into flowers, feeding particularly on pollen grains, and onto leaves, feeding on leaf cells. Some have become predators of small insects and mites. Others moved to feed on fungi under bark and on fungal spores. The key to this diversity of life style is probably a combination of small size and piercing-sucking mouthparts. On the scale at which thrips feed, many potential foods, such as plant cells, are liquids surrounded by a tough protective layer. Thrips' mouthparts are well suited to piercing the tough layer and then sucking out the liquid contents. Thrips have been able to diversify their sources of food and hence their life styles with little or no modification of the mouthparts (Heming, 1993).

Fungus-feeding thrips are described in chapter 2. They exhibit some remarkable behaviours, such as fighting. Thrips on leaves are described in chapter 3. Some are major crop pests, producing feeding damage and spreading viruses. Gall-producing thrips are described in chapter 4. They include the only thrips known to have soldiers. Thrips in flowers are described in chapter 5. These thrips can harm crops through direct damage, but they can also be beneficial as pollinators.

2 Thrips on fungi

People rarely notice fungus thrips, which tend to live concealed under bark or in leaf litter. However, their behaviour is fascinating and they are easy to study.

The thrips that are closely associated with fungi are almost all in the family Phlaeothripidae. This family is divided into two groups (table 2). The majority, about 2,100

Table 2. *Classification of the fungus thrips mentioned in this chapter, and the regions in which they are found. A cross indicates species found in Britain.*

Family	Sub-family	Species	Region
Phlaeothripidae	Phlaeothripinae	*Acanthothrips nodicornis+*	Europe
		Hoplothrips fungi+	Europe
		Hoplothrips ulmi+	Europe
		Hoplothrips karnyi	USA
		Hoplothrips pedicularius+	Europe
	Idolothripinae	*Anactinothrips gustaviae*	Panama
		Elaphrothrips tuberculatus	USA
		Idolothrips spectrum	Australia
		Megathrips lativentris+	Europe

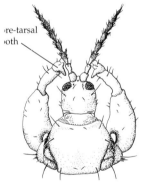

re-tarsal
oth

Fig. 6. An adult male *Hoplothrips pedicularius* with enlarged fore-legs.

Fig. 7. An adult male *Hoplothrips pedicularius* with slender fore-legs.

species, have narrow maxillary stylets (mostly 2–4 µm wide) and are in the sub-family Phlaeothripinae. This sub-family includes species that live on leaves, in flowers and in galls, as well as species that live on fungi. The species that live on fungi appear to feed by extracting juices from fungal hyphae, but virtually nothing is known about how they do it. The remaining 600 or so species have wider maxillary stylets (5–10 µm wide) and are in the sub-family Idolothripinae (Mound & Palmer, 1983). All the idolothripines appear to feed exclusively on fungal spores and their wider maxillary stylets seem to be an adaptation to consuming these spores whole. Their guts often contain intact spores about 4–6 µm wide that can pass through idolothripine stylets, but are too large to pass through the narrow stylets of a phlaeothripine.

There is remarkable variation of body size and shape within species, which makes classification and identification very difficult. Males, in particular, often vary between large forms with grotesquely enlarged fore-legs (fig. 6) and smaller forms with legs as slender as those of the females (fig. 7). Specimens with enlarged fore-legs usually also have a large pointed tooth, known as a fore-tarsal tooth, on the inside of each front foot (fig. 6). Specimens at the two size extremes are termed oedymerous (enlarged) and gynaecoid (female-like). Some also have strange structures on other parts of the body. For example, adults of some species have tubercles (small projections) on each side of the abdomen and some larvae have horns on the head.

The function of these structures has long been a mystery, but recent observations have begun to reveal an advanced social behaviour in fungus thrips and have shown how these strange structures are used. The rest of this chapter describes observations on some of the few species that have been studied. Many other species may behave similarly.

Anactinothrips gustaviae

Many fungus thrips live in large colonies (Ananthakrishnan, 1984a), and *Anactinothrips gustaviae* is one of these. The adult is dark brown or black and about 5 mm long, while the larvae are yellow and 0.5–5 mm long. Groups of 2–200 individuals inhabit permanent communal "bivouac" sites on live tree trunks in Panama (Kiester & Strates, 1984). Adults and larvae form co-ordinated foraging parties that move around on the tree trunk foraging at lichens for 1–6 hours at a time before returning to the bivouac (fig. 8). They lay down chemical trails, which can last 2–3 days, to allow the rest of the group to find food sources. The adults guard the eggs in the bivouac and protect the larvae.

Fig. 8. A foraging group of *Anactinothrips gustaviae* larvae (from a photo by A.R. Kiester).

Hoplothrips pedicularius

Stereum fungus is common on the dead wood of deciduous trees in Britain and it is frequently inhabited by *Hoplothrips pedicularius*. The adult thrips is about 2.5 mm long and is easy to find if you know where to look (p. 55). The behaviour has been studied by Crespi (1986a).

Adult females contribute to a communal egg mass. The adult males use their fore-legs, which vary in degree of enlargement (fig. 6, fig. 7), to fight for possession of an egg mass. A successful male copulates repeatedly with the females at his egg mass and so increases the likelihood that he will be the father of the eggs that are laid. Males copulate more often when the females are laying, which suggests that the last male to mate before egg laying is most likely to fertilise the eggs. The way in which phlaeothripids mate is shown in pl. 2.4.

A male fights by trying to grasp the opponent between its fore-legs and then drive its fore-tarsal teeth into the opponent's body. It often tries to grab its opponent's abdomen from the rear. If both males try to grab each other's abdomen at the same time, they can chase each other round in circles. Most attempts at grasping and stabbing end with the fore-legs slipping off the opponent's body. Occasionally, a male appears to injure its opponent, but no lethal stabbings have been witnessed. Males also wag their abdomens sharply from side to side. This can knock an opponent to one side, and the behaviour appears to be a defence against stabbing from the rear.

Males with larger fore-legs tend to win fights and so obtain more of the matings immediately before eggs are laid. If the males are of similar size, fights can escalate, with each stabbing the other, and can last up to 10 minutes. After a fight, the winner becomes dominant at the egg mass and the loser becomes subordinate. The loser either waits nearby or moves away.

There would appear to be strong selection for enlarged fore-legs. However, the gynaecoid males do not lose out entirely. They can obtain matings by sneaking up to females at egg masses and mating without the dominant male noticing and by mating with females away from egg masses. Males have two alternative tactics for trying to mate with females. Larger males will fight and guard, while smaller males that are unlikely to win fights will try to sneak matings. Males of intermediate size may switch tactics according to circumstances.

Hoplothrips karnyi

The behaviour of *Hoplothrips karnyi* is very similar to that of *Hoplothrips pedicularius*, except that male fighting often results in the death of one of the fighters (Crespi, 1988). The thrips are about 3 mm long and live in colonies on *Polystictus* fungus on trees in the USA. The British species *Hoplothrips fungi* and *Hoplothrips ulmi* are thought to be closely related to *Hoplothrips karnyi*, so they may also have lethal male fighting.

The adult males adopt alternative tactics of fighting or sneaking (fig. 9). Fights last up to 16 minutes and when a male is successfully grasped and stabbed, the victim can be held for several minutes with the fore-tarsal teeth piercing its body.

Males that guard egg masses appear to assess the reproductive condition of the females by embracing them around the abdomen. They then mate more often with the plumper ones that are about to lay eggs.

fighting males

female   sneak male

Fig. 9. A male of *Hoplothrips karnyi* attempting a sneak mating with a female at an egg mass while two other males fight over her (from a photo by B.J. Crespi).

Acanthothrips nodicornis

Larvae of *Acanthothrips nodicornis* have a pair of goat-like horns at the front of the head (fig. 10). The larvae form small groups and hunt for fungi in cracks in tree bark (Pelikan, 1990). When they find some fungal strands, they cluster round with their mouthcones in the crack and their abdomens pointing outwards. They wag their abdomens at each other as if trying to obtain more space. A larva that arrives after the others have started feeding will butt the other larvae aggressively with its horns and try to force its way to the food. Bigger larvae with longer horns have an advantage. This species occurs in Britain, but it is not very common.

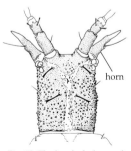

horn

Fig. 10. The head of a larva of *Acanthothrips nodicornis*.

Some other species

Adult males of *Elaphrothrips tuberculatus* also fight with their fore-legs (Crespi, 1986b). The females wag their abdomens to defend their eggs against cannibalism and predation (Crespi, 1990).

Adult males of some species of thrips have been found to fight without using enlarged fore-legs.

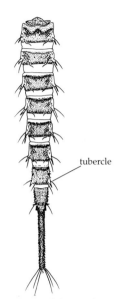

tubercle

The giant Australian thrips, *Idolothrips spectrum*, which can reach a length of 14 mm, does not have fore-tarsal teeth, but has tubercles on the side of the abdomen (fig. 11). Males fight and use their abdomens as weapons (Mound, 1991). *Megathrips lativentris* is about 3.5–5 mm long and is widespread in Britain. The males have tubercles on the side of the abdomen and so it is suspected that they fight in a way similar to that of *Idolothrips spectrum*. This possibility would be worth investigating. However, they also have a large pair of horns on the sixth segment of the abdomen (fig. 12). These horns stick out sideways in specimens mounted on microscope slides, but they are upright in live specimens. They are hollow and there is an opening at the tip, so perhaps they are used to release a pheromone (Pelikan, 1994). Nothing is known about the behaviour of this species.

Fig. 11. The abdomen of an adult male *Idolothrips spectrum*.

horn

Fig. 12. The abdomen of an adult male *Megathrips lativentris*.

pheromone: a chemical message between members of the same species

3 Thrips on leaves

Some of the most damaging thrips pests live on leaves. When leaf thrips become abundant on a plant, their feeding and egg laying can cause severe damage to leaf tissue and even lead to complete defoliation. Some thrips, such as *Thrips tabaci*, can pick up a plant virus by feeding on an infected plant and then spread the virus later on by feeding on uninfected plants. Thrips that spread viruses can do a lot of damage, even at very low infestation levels.

Both adults and larvae of leaf thrips feed by sucking the juices from individual leaf cells (Chisholm & Lewis, 1984; Heming, 1993). The mouthcone is applied to the leaf surface (fig. 13). The mandible is pushed down so that it pierces a leaf cell wall and then the maxillary stylets are inserted in a series of rapid thrusts. The liquid contents of the cell are sucked up through the stylets. After a period ranging from a few seconds to 30 minutes, the mouthparts are withdrawn and the thrips moves on, possibly to feed straight away at another site nearby. Feeding marks can be seen when the surface of the leaf is examined with a scanning electron microscope (fig. 14). Where the mouthcone was pressed against the leaf, the wax has been removed from the leaf surface, leaving a round mark. In the middle of the mark is a figure-of-eight hole where the mandible and stylets pierced the surface. Severe feeding damage is often apparent to the naked eye as a silvering of the leaf surface where cells have been emptied and there can be yellowing or browning of leaves where cells have died around the injury site. Cabbage thrips (*Thrips angusticeps*) feed on young leaves, which become stunted and turn yellow where the thrips have fed (pls. 1.5–1.7).

Fig. 13. An adult *Limothrips cerealium* feeding on a leaf (from a photo by I.F. Chisholm).

Insects that live on plants must overcome a number of difficulties associated with their way of life. Leaf surfaces can be very smooth and thrips could easily lose their grip. The arolium on the end of the leg (see chapter 1) helps thrips grip surfaces. Many thrips retreat into small crevices, such as between leaves. Here they are less likely to be blown or shaken off and they may also find themselves in a more humid microclimate where they are less likely to be desiccated (Unwin & Corbet, 1991). Thrips migrate regularly to new hosts. Experiments can be carried out with coloured and scented traps to investigate the responses of thrips during flight. Leaf thrips seem to respond differently from flower thrips (Kirk 1984a, 1985a). When some species that live on the leaves of grasses were compared with some that live in flowers, catches of leaf thrips were relatively little affected by flower colours and a flower scent, whereas the flower thrips responded strongly to both. It is not known whether leaf thrips respond in flight to leaf scents.

Fig. 14. A feeding mark about 20 μm across on a wheat leaf (from a photo by I.F. Chisholm).

The rest of this chapter describes some aspects of the biology of one of the most studied leaf thrips.

Fig. 15. An adult female
Limothrips cerealium.

The grain thrips (*Limothrips cerealium*)

This species breeds on grasses and cereal crops, particularly wheat. It is common across western Europe, including Britain, as well as in many other temperate regions around the world. The adult female is black and about 1.7 mm long with a characteristic elongate head and triangular end of the abdomen that allow it to be recognised with the naked eye (fig. 15). The adult male is also black, but smaller, and it has no wings, so if you recognise a *Limothrips cerealium* that has landed on you, you also know that it is a female. The larvae are pale yellow.

The life cycle has been studied in southern England (Lewis, 1959). Adult females spend the winter in plant debris and on or under bark. In April, they migrate to grasses and winter cereals and start laying eggs in the leaf bases and leaf sheaths. As the plant grows, the thrips move up towards the ear. Sometimes the adults will fly on to lusher plants, such as spring wheat. The larvae feed on the leaves and pupate on the plant. The life cycle takes about 4–5 weeks from egg to adult. Adult males and females of the new generation appear in June and July. They mate on the plants and the wingless males die soon afterwards. As cereals start to dry out in July and August, the adult females migrate to overwintering sites, often in vast numbers. There appears to be only one generation per year in southern England.

Although these thrips may reach very high densities in cereal crops, there is usually little visible damage from feeding or egg laying. However, there does appear to be a loss of yield. One insecticide advertisement has described cereal thrips as "the enemy nobody suspected" because of the subtle way in which they reduce yield. In experiments performed by Chisholm & Lewis (1984), developing ears of wheat with 20 adult thrips caged on them produced 42% fewer grains than ears without thrips.

Cereal thrips certainly feed on leaf cells, but do they also feed on pollen? Grasses and cereals are wind-pollinated and produce large amounts of pollen which is blown on the wind and settles all over plants. It is not known whether cereal thrips can feed on pollen in the same way as flower thrips (chapter 5). This could be investigated easily (p. 56). Pollen, which is highly nutritious, might even form a large part of the diet when grasses and cereals are in flower.

Adults of *Limothrips cerealium* have stout spines and tubercles near the end of the abdomen. Males have two pairs of tubercles, each with stout spines on the end, on the upper surface of the ninth abdominal segment (fig. 16) and females have a pair of stout spines on the upper surface of the tenth abdominal segment (fig. 17). Do they use these structures for fighting? If so, what resource could they be defending? Patient observations could well supply the answers to these questions.

Adult females migrate from cereal fields in July and August when the plant water content drops below about 45%, possibly because it is then difficult for them to feed. The

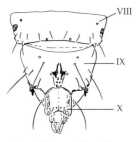

Fig. 16. The tip of the abdomen of an adult male *Limothrips cerealium*.

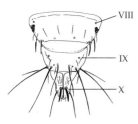

Fig. 17. The tip of the abdomen of an adult female *Limothrips cerealium*.

thrips fly by day when the air temperature is above 20 °C. When temperatures do not rise above this threshold for several days, large numbers of thrips may reach a stage at which they are ready to migrate. Mass flights then occur when the temperature eventually crosses the 20 °C threshold (Lewis, 1964). Thrips can be seen drifting in the air, but they only become obvious when they settle.

Limothrips cerealium and other cereal thrips are popularly believed to fly in thundery weather, hence the popular name of "thunderbugs" or "thunderflies" for cereal thrips in general. Although experiments with water traps have not found any clear relationship between thunder and mass flights, there may still be some link with thunder. The occasion when I was most troubled by thrips landing on me was five minutes before a very large thunderstorm. There were so many cereal thrips landing and causing itching skin that people were rushing indoors to get away from them. The itching is probably caused by thrips probing skin as if they were probing a leaf.

A migrating thrips seeks out a small crevice in which to spend the winter. It prefers a crevice about 300 μm wide, but thrips manage to penetrate gaps that are as little as 50 μm wide. The crevice can sometimes be the gap beneath the glass of a picture frame. When the chosen crevice is inside a smoke detector, the thrips may set off the detector and cause considerable inconvenience. Public buildings, such as hospitals, may have to be evacuated and the fire brigade may be called out. In Suffolk, 25% of false alarm calls to the fire brigade in July are thought to be triggered by insects. Most of these insects are probably thrips because the false alarms tend to be on the days when mass flights would be predicted from the weather conditions (Cuthbertson, 1989).

4 Thrips in galls

A gall is an abnormal plant growth caused by an insect, a mite or a disease (Redfern & Askew, 1992). Over 300 species of thrips cause galls and nearly all of them are in the sub-family Phlaeothripinae of the family Phlaeothripidae. Gall thrips account for only about 7% of the known species of thrips, but these few species show some of the most remarkable of all thrips behaviours. Unfortunately, there are no gall thrips in Britain. They are found mainly in India, southeast Asia and Australia.

Thrips form galls by feeding on very young leaves and thus modifying their growth. Galls can take the form of simple curls, rolls and folds of leaves, or more complicated forms, such as rosettes, pouches and horns (fig. 18) (Lewis, 1973; Ananthakrishnan, 1984a, 1984b). The gall is usually completed within a few days of a thrips starting to feed and it then provides a sheltered environment within which the thrips can feed and rear its young. In some galls, the thrips are sealed in and cannot leave until the gall splits open. Galls are normally started by a single adult female thrips that can lay hundreds of eggs. The thrips can breed in the gall for one or more generations. Populations of hundreds to thousands of individuals can build up in a single gall that might be only a few centimetres across.

Fig. 18. Horn galls on a leaf.

Life is tough for gall thrips. They are attacked in their home by a range of insects. As soon as a foundress thrips has formed a gall, other members of the same species will try to take it over because it is quicker to use a ready-made gall than to go to the trouble of making one. Predatory inquiline thrips will try to invade the gall and eat the foundress and her young. A range of predators and endoparasitoids are also likely to attack gall thrips within their gall.

inquiline: an animal that lives in the home of another species

endoparasitoid: an insect that develops parasitically inside the body of another insect

A gall offers some protection to the occupant, but it is a valuable resource that needs to be defended against intruders. It also advertises the presence of the occupants to predators and endoparasitoids.

The predatory inquiline thrips *Androthrips flavipes* takes advantage of the galls of many species of thrips in India (Ananthakrishnan, 1984b). The adult female enters the gall, lays her eggs among those of the host, and eats host eggs and larvae. The inquiline eggs hatch after a few days and the inquiline larvae also eat the eggs and larvae of the host.

Gall thrips commonly have enlarged fore-legs and fore-tarsal teeth in the females, but not the males. In contrast, fungus-feeding thrips have enlarged fore-legs in the males, but not the females (see chapter 2). In each type of thrips, the sex with enlarged fore-legs is the one that defends a resource. Male fungus thrips defend egg-laying females or egg patches, whereas female gall thrips defend the galls they have formed. We would expect to see female gall thrips using their fore-legs for fighting in the same way as male fungus thrips.

Fig. 19. An adult female *Kladothrips rugosus* with enlarged fore-legs.

phyllode: a leaf-like flattened stem

Crespi (1992a) has discovered fighting in several species of gall thrips. Adult female *Kladothrips rugosus* are about 2.5 mm long. They form spherical pouch galls about 8 mm across on the phyllodes of *Acacia pendula* in Australia. The females fight over galls, often to the death. They face each other with their enlarged fore-legs held out (fig. 19) and rear up at an angle of 30–40° so that they are touching each other with the bases of their antennae. Then they sway from side to side, stay still, and lunge forward, each attempting to grasp and stab the opponent's head or thorax with her fore-legs. Fights usually break up after about a minute unless one of the thrips has a firm grasp, in which case the victim is held aloft so that its feet are off the ground. The victim cannot struggle free because it is pierced on each side by the victor's fore-tarsal teeth. The victor moves its fore-legs one at a time every few seconds, stabbing the victim repeatedly with its fore-tarsal teeth for up to 50 minutes. The victim oozes blood and can be partially crushed into a misshapen mass. It is put down after it has died. Such violence between members of the same species is rare among animals. Galls must be extremely valuable to gall thrips!

In contrast to this violence, we might also expect to see some form of social co-operation in gall thrips, because they possess several of the characteristics that are thought to promote it, including generation overlap, having a shared home to defend, and being haplodiploid (see chapter 1). The connection between haplodiploidy and sociality is not immediately obvious. Haplodiploidy has the curious genetic effect that a female has more genes in common with her sisters than with her daughters (Wilson, 1971), so a female might be able to pass on more of her genes by protecting her sisters than by rearing her own children. We see this kind of behaviour in worker bees and soldier ants. Workers and soldiers are special forms, known as castes, that are adapted to care for the colony or defend it against attackers, rather than rear their own offspring. When haplodiploidy is combined with living in a large colony of relatives that have a shared home to defend, as in gall thrips, the evolution of castes becomes more likely.

The Hymenoptera (ants, bees and wasps) is one of the few insect orders that is haplodiploid, and castes have evolved independently many times within it (Wilson, 1971). However, it could be that some other factor shared by Hymenoptera, such as having a sting, could be responsible rather than haplodiploidy. The discovery of castes in another haplodiploid order would support the theory that haplodiploidy is a contributory factor.

Soldier thrips, that defend rather than reproduce, have been discovered recently by Crespi (1992a, 1992b) in the gall thrips *Oncothrips tepperi* in Australia. The adults are about 3 mm long and form semi-oval galls, 4–11 mm long by 2–5 mm across, on the phyllodes of several species of *Acacia* (fig. 20). Both sexes show a range of fore-leg enlargement with fore-tarsal teeth, but this is greater in females than in males.

Fig. 20. Galls formed by *Oncothrips tepperi* on *Acacia* phyllodes (from a photo by B.J. Crespi).

Fig. 21. A soldier of
Oncothrips tepperi.

Fig. 22. A wingless soldier of
Oncothrips habrus grasping a
winged *Koptothrips dyskritus*
(from a video by B.J. Crespi).

An adult female flies to a phyllode and forms a gall.
Until the gall closes up, she defends it against other adult
females of the same species. When an attacking female
approaches, the owner may run up to her. The fighting
behaviour is similar to that of *Kladothrips rugosus*. The
females rear up head to head with their enlarged fore-legs
held out and attempt to lunge forward and grasp the
opponent's body between their fore-legs and stab it with
their fore-tarsal teeth. They grapple with each other,
alternately lunging forward and then separating. When an
opponent is grasped successfully, she is sometimes lifted off
the ground and held in the air. Crespi (1992a) observed one
lift lasting 51 minutes; the victim died an hour after being let
go. Stabbed females depart, sometimes fatally wounded.

The first generation of offspring within galls of
Oncothrips tepperi contains both winged forms and short-
winged forms. The short-winged forms emerge first (fig. 21).
They cannot fly. They have larger fore-legs and smaller
developing eggs than their mother. This suggests that they
are adapted morphologically for defence, rather than
colonisation and reproduction, and so can be described as
soldier thrips.

Soldiers are needed to defend the gall against attack.
The defensive behaviour of the soldiers can be demonstrated
experimentally. When a hole is poked in a gall, several
soldiers will rush to the hole and look out. *Koptothrips
flavicornis* is a common predatory inquiline that often
manages to take over the galls. If a *Koptothrips flavicornis* is
held near the hole, the soldiers will make an unprovoked
attack on it, using their fore-legs to grab it and stab it.
Soldiers of *Oncothrips habrus* behave in the same way when
Koptothrips dyskritus approaches (fig. 22). The soldiers will
also attack caterpillars and ants, but not members of their
own species. This may prevent them from accidentally
attacking members of their own gall.

5 Thrips in flowers

On a fine summer day it is easy to find flowers with many thrips in them. Flower thrips are abundant. They occur in all three of the main families of thrips, but most are in the family Thripidae.

It is not easy for thrips to live in flowers, because most flowers last only a few days. Flower thrips frequently have to find new flowers. This is particularly difficult for larvae, which cannot fly, and is impossible for eggs, propupae and pupae. Flower thrips usually minimise the time spent in flowers by laying eggs in stems outside the flower, shortening the larval stages to only a few days or weeks, and spending the propupal and pupal stages on or in the soil below the host-plant. The compensation for the difficulty of living in flowers is that flower thrips can feed on pollen, which is rich in protein. This is good for larval growth and the production of eggs. Flower thrips also feed on nectar and floral tissues. Occasionally they can also be predators of mites and other thrips.

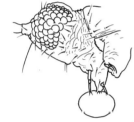

Pollen grains are typically about 25–40 μm across and consist of a tough outer shell with liquid contents. The maxillary stylets through which flower thrips suck up their food are only about 1–2 μm across, so flower thrips cannot swallow pollen grains whole; the liquid contents of the grain must be extracted. The feeding behaviour can be observed easily (techniques p. 56). Larvae and adults approach a grain, place the tip of their mouthcone over it and immediately suck out the contents (fig. 23). The mandible pierces the shell and the maxillary stylets are inserted to suck out the liquid. Pumping muscles can be seen pulsating within the head. The grain usually buckles inwards and crumples up as it is emptied (fig. 24). The thrips nods its mouthcone backwards, dislodging the grain, and then moves on. The whole process usually takes only a few seconds (Kirk, 1984b, 1987). A thrips can eat hundreds of grains per day.

Fig. 23. Head of a thrips feeding on a pollen grain. The part of the front leg that would obstruct the view has been omitted.

A stain can be used to distinguish empty grains from intact grains. Cotton blue in lactophenol leaves empty grains clear, but stains intact grains blue. Experiments using this stain have shown that daily consumption rates per thrips are about 0.5% of the pollen production of an average flower (Kirk, 1987).

Flower thrips have been regularly ignored by pollination biologists. However, much of the pollen in infested flowers can be destroyed by thrips, so thrips could reduce the amount of pollination. Thrips are covered with pollen while feeding and adults regularly move between flowers, so they might also pollinate. Very little is known about the amount of harm and benefit that flower thrips do in flowers.

Fig. 24. A pollen grain buckled after being fed on by a thrips.

The abundance of flower thrips makes them particularly useful for studies of population dynamics. The plague thrips, *Thrips imaginis*, is common in flowers of a wide range of plants in Australia. In some years there are massive outbreaks that cause considerable damage to apple blossom in orchards and even drive people off beaches. Davidson & Andrewartha (1948) studied the population size in Adelaide, South Australia, from 1932 to 1946 by counting the number of adults in 20 roses each day. By the end of the study, they were able to predict the density of thrips at apple blossom time from the temperatures during the previous autumn and winter, and their method was used for many years as an early-warning service for apple growers. The findings were discussed in an ecology textbook by Andrewartha & Birch (1954) and a major controversy ensued over the nature of the factors that determine population size. Flower thrips acquired a certain notoriety among ecologists. The controversy is still discussed in ecology textbooks today (Begon and others, 1990).

The rest of this chapter describes various aspects of the biology of four common British flower thrips. All the species are easy to find and observe, and all need further study.

The rubus thrips (*Thrips major*)

This is one of the commonest thrips throughout most of Britain. It is unusual because it is found in flowers of a wide range of species. A few other species of British flower thrips are similar (table 3), but most are highly host-specific, being restricted to the flowers of only a few related plant species. The easiest way to find the rubus thrips is to look in the large white flowers of hedge bindweed (*Calystegia sepium*) or great bindweed (*Calystegia silvatica*) on warm sunny days. Bindweed is common in country hedgerows as well as on derelict sites in cities. The small yellow males

Table 3. *Classification of some common flower thrips. An asterisk indicates species that are found in flowers of a wide range of species. A cross indicates species found in Britain. See Mound and others (1976) for details of the many rarer and more host-specific species.*

Family	Species	Common name
Aeolothripidae	*Aeolothrips intermedius**+	–
Thripidae	*Ceratothrips ericae*+	–
	Frankliniella intonsa*+	–
	Frankliniella occidentalis*+	western flower thrips
	Kakothrips pisivorus+	pea thrips
	Odontothrips ulicis+	gorse thrips
	Thrips atratus*+	carnation thrips
	Thrips flavus*+	honeysuckle thrips
	Thrips fuscipennis*+	rose thrips
	Thrips imaginis*	plague thrips
	Thrips major*+	rubus thrips
	Thrips vulgatissimus*+	–
Phlaeothripidae	*Haplothrips leucanthemi*+	ox-eye daisy thrips
	Haplothrips niger	red clover thrips

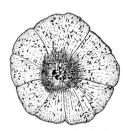

Fig. 25. A swarm of male thrips in a flower of *Calystegia sepium*.

often swarm over the flowers (fig. 25), along with the similar males of *Thrips fuscipennis*. The mating behaviour can be easily observed in a flower (Kirk, 1985b).

The males patrol the flower, often bumping into each other. The females, which are larger and darker than the males, occasionally land on the flower. When a female lands, the nearest males will climb all over her, attempting to mate. A frenzied ball of males can form around the female for a while. Sometimes the female loses her grip and the whole group rolls off the flower. Eventually one male is successful and the others leave. Copulation lasts several minutes, after which the female flies away.

Experiments with coloured and scented water traps (Kirk, 1984a, 1985a) have shown that *Thrips major* adults particularly land on traps painted white or scented with anisaldehyde, a widespread flower scent. Flower thrips can distinguish colours and scents while in flight and actively control where they land, so their flight is not as weak and passive as their small size would suggest.

The western flower thrips (*Frankliniella occidentalis*)

This species (pl. 1.1) is common in the western USA. During the 1980s it spread across Europe and it was first found in Britain in 1986. The spread has been mainly the result of the long-distance trade in glasshouse plants. This thrips is now a major pest of horticultural glasshouse plants, such as African violets and pot chrysanthemums. In Britain it is common in glasshouses, but it does not seem to survive outdoors over winter.

When it feeds on petals, it removes the pigment and produces dark excretory droplets (pls. 1.2,1.3). Infested plants are unsightly and so difficult to sell. The thrips breed fast under hot humid glasshouse conditions, going from egg to adult in 2–3 weeks. They are difficult to treat with chemicals because they are usually deep within the flowers and are resistant to some insecticides (Helyer & Brobyn, 1992). The practicality of biological control in glasshouses is being investigated using a range of organisms: predatory *Orius* bugs, predatory *Amblyseius* mites, *Ceranisus* endoparasitoid wasps, and insect-killing *Verticillium* fungi (Parker and others, 1995). The thrips also cause considerable damage by spreading tomato spotted wilt virus (TSWV) on a range of plants, not just tomatoes (pl. 1.4). Virus particles are picked up when a thrips larva feeds on the contents of infected plant cells. The virus replicates in the thrips and then is later injected along with saliva whenever the larva or subsequent adult feeds. Curiously, adults cannot pick up the virus and become infective. The western flower thrips is not entirely harmful. In cotton crops, it can be a useful predator of spider mites.

Swarms of males have been observed on large white surfaces during hot humid weather conditions in Arizona (Terry & Gardner, 1990). The behaviour is similar to that of *Thrips major*, except that the males sometimes fight each

endoparasitoid: an insect that develops parasitically within the body of another insect

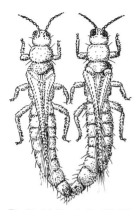

Fig. 26. Adult male *Frankliniella occidentalis* fighting with their abdomens.

other. Two males line up alongside each other, facing the same direction, and wag their abdomens from side to side two to three times per second (fig. 26). They hit each other with their abdomens and sometimes one manages to flick his opponent over. The fights last about 15 seconds on average. The males do not have enlarged fore-legs or tarsal teeth, so there is no stabbing.

This behaviour is curious because the combatants do not appear to gain anything. Females are not nearby and the fighting males seem to be less likely to encounter females than the patrolling males. The behaviour might be useful in small flowers if it established exclusive mating on a whole flower on which females were particularly likely to land. It would be interesting to know more about this behaviour and whether it occurs in many other flower thrips. Fighting between males of *Frankliniella intonsa* has been observed in flowers of field bindweed (*Convolvulus arvensis*) in full sun in southern Germany. Fighting has also been observed between males of *Aeolothrips intermedius* in Germany. Both thrips are common in Britain, so it should be possible to observe them fighting here.

Some thripids and phlaeothripids produce droplets at the end of the abdomen in response to prodding or predator attack (Lewis, 1973). The liquid produced by larvae of the western flower thrips is a mixture of two chemicals, decyl acetate and dodecyl acetate (Teerling and others, 1993). Larvae and adults walk away from it or even drop off the leaf they are on. The scent seems to be acting as an alarm pheromone, warning other thrips of danger. It may be possible to use this alarm pheromone to protect plants, possibly in conjunction with other measures.

Experiments with coloured traps have shown that the western flower thrips is particularly caught by certain shades of blue (Matteson & Terry, 1992). Blue sticky traps are used in glasshouses to monitor western flower thrips without catching the beneficial predators and endoparasitoids that are used for biological control.

The pea thrips (*Kakothrips pisivorus*)

This species breeds in flowers of peas, beans and vetches, and heavy infestations can damage the flowers. When the flowers have dropped off, larvae on peas (but not beans) continue feeding on the developing pods and damage these as well (pl. 1.9). In Britain, the pea thrips is common where peas and beans are grown regularly. It has not been recorded from Scotland. The species is often referred to under the alternative name of *Kakothrips robustus*.

The adults are a shiny black and the larvae are yellow with a black tip to the abdomen (pl. 1.8). This makes them easily distinguishable with the naked eye from other adults and larvae in the flowers. Behavioural observations can be made with confidence about which species is involved.

The life cycle in southern England is described by Williams (1915). There is probably only one generation per year. In May, the adult males and females emerge, fly to the flowers and mate. Eggs are laid in the stamens and hatch after 7–10 days. First-stage larvae feed for 8–9 days and then moult. Second-stage larvae feed in the flower for about 6 days and then drop to the ground. They bury themselves up to 30 cm below the soil surface, where they stay until the following March to May, and then moult to a propupa and then a pupa, before emerging from the soil. The reason for over-wintering as a larva rather than a pupa is unclear.

It is unusual for flower thrips to lay their eggs in flowers because this gives little time for the larvae to develop before the flower wilts. As a result, the larvae are more likely to have to undertake a potentially risky journey to find a new flower.

The movement of larvae between flowers of the field bean (*Vicia faba*) has been studied by Kirk (1985c). The eggs and larvae need to spend about 24 days in flowers, whereas each flower only lasts about 8 days. Since the larvae are restricted to the flowers when on field beans, each has to move to a new flower at least twice. This is usually easy because flowers open in succession along a raceme and a larva leaving a wilting flower can usually walk straight into an adjacent younger flower. However, when the last flower on a raceme begins to wilt, the nearest young flower is on the next raceme up the plant. Larvae accumulate in the last flower on a raceme. It is not known whether the young larvae in the last flower manage to move to a different raceme.

Adult male *Kakothrips pisivorus* have tubercles on the side of the abdomen (fig. 27), which look like small versions of those found in the fungus thrips *Idolothrips spectrum* (chapter 2, fig. 11). It would be interesting to know whether the males fight with their abdomens in a similar manner.

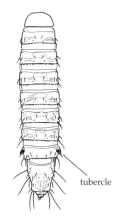

tubercle

Fig. 27. The abdomen of an adult male *Kakothrips pisivorus*.

The ox-eye daisy thrips (*Haplothrips leucanthemi*)

The ox-eye daisy thrips occurs throughout Britain. It breeds on established patches of ox-eye daisy (*Leucanthemum vulgare*) by the sides of roads or on wasteland. New patches, such as those sown recently with wild flower seed, may take a few years to be colonised. The species is easy to find because the black adults are often clearly visible on the flowers (pl. 2.8). Note the confusing similarity of the names of the genera *Haplothrips* (flower thrips) and *Hoplothrips* (fungus thrips).

genera: groups of related species

Very little is known about the biology of this species. *Haplothrips niger*, which is common on clover in the USA and Canada, is very closely related and may even be a strain of the same species, so details of its life cycle may be similar (Loan & Holdaway, 1955).

During just three hours of observation of a few adults in a flower, I saw adults feeding (pls. 2.1, 2.2), grooming (pl. 2.3), mating (pl. 2.4) and laying eggs (pl. 2.5). Longer sequences of observations could reveal details of the whole life cycle.

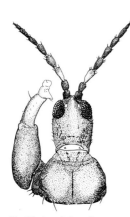

Fig. 28. An adult male
Haplothrips leucanthemi with
enlarged fore-legs.

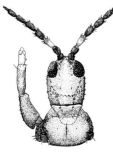

Fig. 29. An adult male
Haplothrips leucanthemi with
slender fore-legs.

The florets within a daisy flower are youngest in the middle and oldest at the edge. Florets within a ring are of similar age. Each floret releases pollen first and then later projects its stigma so that it can receive pollen. A typical flower will therefore have an inner ring of young florets releasing fresh pollen and an outer ring of older florets with projecting stigmas. Adults feed at the inner ring, but lay eggs in the outer ring. Female thrips covered in pollen can be observed walking away from the feeding site, laying an egg within the older florets that are ready to receive pollen, and then returning to the feeding site. Is this an adaptation to bring about pollination? If so, what is the advantage to the thrips? Could it be a behaviour to avoid cannibalism, by keeping eggs well away from the feeding site? This could be investigated by recording the survival of eggs moved to florets of various ages.

The adults have a defence against predators, which can be demonstrated artificially. When an adult is prodded with a pin, it quickly swings its abdomen towards the pin and deposits a drop of liquid on it. The liquid is a volatile substance called mellein (Blum and others, 1992). Experiments have shown that it repels ants, which are likely predators of thrips.

Some males have enlarged fore-legs with fore-tarsal teeth (fig. 28, fig. 29). Fighting has never been observed in this species, but probably no one has ever looked for it! What could the males be fighting over? Adult females seem to be relatively immobile, not flying away when disturbed. Males could be defending a harem of females within a flower, ensuring that only they are the fathers of any eggs that are laid. It would be interesting to see what happens if several males with enlarged fore-legs are placed in the same flower. The males with enlarged fore-legs may be fighters, while the smaller males may be adopting the alternative tactic of trying to sneak matings without the fighters noticing, as in the fungus thrips *Hoplothrips karnyi* (chapter 2).

6 Identification

Introduction to the keys

These keys can be used to identify adults, pupae, propupae and larvae of British thrips. They are not intended to be comprehensive. In most cases, the commoner species are keyed out to species and the rarer species are keyed out only to genus (a group of one or more species) or to family (a group of one or more genera). Specialist keys to adults (Mound and others, 1976) or larvae (Speyer & Parr, 1941) can be used to identify the rarer species further.

The keys cover the 158 British species listed by Mound and others (1976), as well as the western flower thrips (*Frankliniella occidentalis*), a pest species which has arrived in Britain recently. The species names given by Mound and others (1976) are used throughout. Commonly-used earlier names and some recently changed names are also given.

Adults have been well collected and studied in the past and so can all be identified to species, at least in Britain. In contrast, immature stages are poorly known, partly because they have fewer features that differ between species, so species identification is not always possible (Heming, 1991). Some larvae are identified to species in these keys, but specimens should be checked against the more detailed descriptions of thripid and aeolothripid larvae in Speyer & Parr (1941) for confirmation. Pupae are even more poorly known than larvae (Heming, 1991). They have even fewer diagnostic features and are rarely collected because they are usually in soil and so are hard to find. Pupae are identified only to family and instar in these keys. There are no other keys for further identification of pupae.

Before using the identification keys, you need to be sure that the specimen you wish to identify is really a thrips! As a quick guide, compare it with the thrips in pl. 3 and pl. 4. If it roughly resembles any of them, check the following characters.

1) Body length, excluding antennae and bristles, not more than 7 mm, and usually not more than 2 mm. This applies to British thrips. Thrips from some other countries can reach a massive 14 mm!

2) Six legs.

instar: the form of a developmental stage between moults

3) Each leg tipped with a rounded, expandable bladder (arolium) that is about as wide as or wider than the leg (fig. 30). The bladders may have collapsed in specimens on microscope slides.

Fig. 30. Tip of fore-leg with expanded arolium on the end.

Fig. 31. Mouthcone viewed from the side.

4) Head with a cone-shaped structure (mouthcone) underneath, projecting downwards and slightly backwards (fig. 31).

5) Needle-like, sucking (not biting), mouthparts. These are usually retracted within the mouthcone, so they are not normally visible unless specimens are mounted on microscope slides (fig. 32). The bugs (Order: Hemiptera) also have needle-like, sucking mouthparts, but these rest exposed, often against the body.

If your specimen fits this description, it is a thrips adult or larva. Thrips pupae lack the expandable bladders on the feet and the mouthparts are not developed, so they have no clear diagnostic features. They can be recognised by their general appearance, but this requires experience! Thrips are most likely to be confused with springtails, small staphylinid beetle adults or cecidomyiid fly larvae. If your specimen is not a thrips, then an insect/invertebrate identification book, such as Chinery (1976, 1986) or Tilling (1987) may help you to find out what it is instead. The narrow, strap-like, fringed wings that are characteristic of most adult thrips are not included among the five recognition characters given above because such wings are not present in immature thrips or in wingless adults. Some very small insects in other orders also have narrow, fringed wings.

needle-like mouthparts

Fig. 32. Head and mouthcone as seen in a specimen mounted on a microscope slide.

When a pupa is about to moult to an adult, the adult features show through the pupal skin. This is confusing for users of identification keys. Specimens that appear to have an inner and an outer skin should be avoided. If you rear thrips, you can use these keys to identify pupal instars.

It takes time to work through the above characters to check that each specimen is a thrips, but beginners should check them all carefully at first. With experience, it becomes possible to recognise both immature and adult thrips easily by their general form with only the naked eye or a hand lens, without having to check all these characters. If thrips are observed carefully when alive and then mounted and identified, one can learn to recognise some distinctive species alive in the field from just their colour, shape and behaviour, and this can be invaluable for ecological and behavioural studies. The species of plant on which thrips are found can give useful clues about the identity of thrips. Table 4 (p. 61) lists some common plants and the thrips species that are particularly associated with them.

Before using the keys, mount your specimens on microscope slides and label them (technique p. 57). The mounted specimens should be viewed with a compound microscope that can magnify at least x400 (for example a x10 eyepiece with x40 objective). For identification of some larvae to species, even higher magnification (x1000 with oil immersion) is needed. If phase contrast is available, it should be used, as it makes many features show up more clearly. By focusing carefully, it should be possible to move between views of the upper surface and the lower surface of the specimen. Check which way the focusing knob moves the objective lens so that you can easily tell which surface of the thrips is which.

To identify a specimen, start at Key I. The key consists of a series of numbered couplets, which are pairs of contrasting descriptions. Start at couplet 1 and decide which of the two descriptions fits your specimen. At the end of the description that fits, you will find the number of the couplet or key to go to next, or the name of the species or group to which your specimen belongs. If neither description in a couplet fits, you have probably made a mistake. Start again, taking care over difficult couplets. Consultation with a colleague or teacher may help.

Very few species of thrips have common names. The key gives the full scientific names. A full scientific species name is usually written in three parts, for example *Thrips tabaci* Lindeman. The first word, by convention printed in italics with a capital letter, is the genus, the group to which the species belongs. The second word, in italics without a capital letter, is the specific epithet. The third word, not in italics, is the name of the author who first published a description of the species. This is often omitted. Well-known authors' names are often abbreviated. If the name of the genus has been changed since the author named the species, the author's name will appear in brackets. When the genus and specific epithet cannot be italicised, as for example when handwritten, they are underlined instead.

In order to identify most specimens, it is necessary to calculate ratios of lengths of different parts of the specimen. This can be done accurately by means of a graticule (a glass disc with a miniature scale marked on it) in the eyepiece of the microscope. For a few species in Key I and Key III and all species in Key IV, it is necessary to measure lengths in micrometres. For this, the graticule must be calibrated against a stage micrometer (a microscopic scale marked on a microscope slide). If a microscope has a drawing attachment, then it can be used in conjunction with a stage micrometer to make accurate measurements. Body lengths given in the key are the lengths of mounted specimens, excluding antennae and bristles. Some individuals may be unusually large or small, so the sizes should be used only as a guide unless given first in a couplet as the main diagnostic character. Lengths of segments are measured along the mid-line of the body and widths of segments are measured across the body. Lengths of setae (bristles) are measured along the setae. For curved setae, it may be necessary to divide them by eye into two or three roughly straight sections, measure each section and then add the lengths of these together.

Particular segments of the abdomen or antennae, as well as structures that are repeated on each segment, are specified by Roman numerals. The segments are numbered in the order in which they occur, counting away from the head (fig. 33).

With these keys, it should be possible to identify most specimens unless they are damaged or poorly mounted. If body contents or dark pigments have not been removed during the preparation of the specimen, it may not be possible to see some important features. If specimens are not straight or are squashed, measurements of lengths may be misleading.

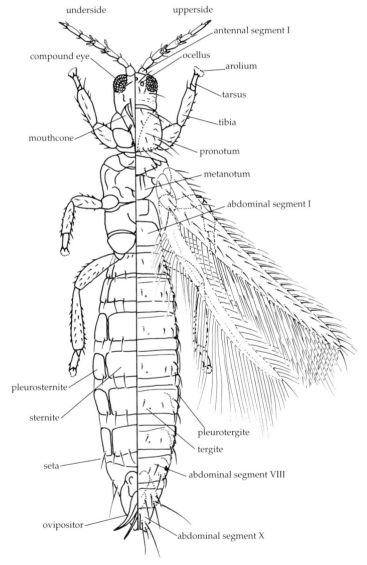

Fig. 33. A diagram of an adult female thripid. The left half shows the underside and the right half shows the upperside.

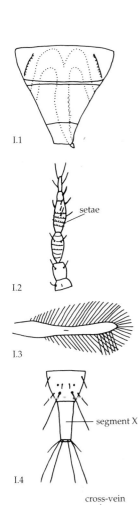

I The families and immature stages of thrips

The bristles on thrips are called setae (fig. 33). For some species, it is useful to measure the lengths of setae in micrometres, so you may need an eyepiece graticule that has been calibrated against a stage micrometer.

1 With a Y-shaped ovipositor (egg-laying apparatus) in segments VIII–X of the abdomen (fig. 33, I.1).
        Adult female 2

– Without a Y-shaped ovipositor in segments VIII–X of the abdomen   4

2 Antenna with 6, 7 or 8 segments (examine the tip closely for small segments)   Family THRIPIDAE Key III

– Antenna with 9 segments   3

3 Forewings long, reaching beyond the tergite (upperside plate, fig. 33) of abdominal segment II   8

– Forewings absent or short, not reaching beyond the tergite of abdominal segment II   35

4 Antennae distinctly segmented, projecting forwards (pls. 3.2, 3.3, 4.2, 4.3), longer than the maximum width of the head, often with rows of setae (bristles) or of short spines on at least some segments (I.2)   Larva or adult 5

– Antennae indistinctly segmented, projecting forwards, sideways or backwards (pls. 3.4, 4.1), sometimes shorter than the maximum width of the head, never with rows of setae or of short spines on the antennae
        Propupa or pupa 37

5 Forewings long, reaching beyond the tergite (upperside plate, fig. 33) of abdominal segment II   6

– Forewings absent or short, not reaching beyond the tergite of abdominal segment II   9

6 Forewings without setae in the outer half (I.3) (ignore the long hairs that fringe the wing); abdominal segment X tubular and as if cut off at the tip (truncate) (I.4)
Adult male or female, family PHLAEOTHRIPIDAE Key IV

– Forewings with setae in the outer half (I.5, I.6); abdominal segment X bluntly rounded (I.7)
        Adult male 7

7 Wings broad, with 3 or 4 visible cross-veins (I.5); antenna with 9 segments (examine the tip closely for small segments)   Family AEOLOTHRIPIDAE Key II

– Wings narrow, with at most 1 visible cross-vein (I.6); antenna with 6–9 segments
        Family THRIPIDAE Key III

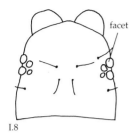

I.8

8   Wings broad, with 3 or 4 visible cross-veins (I.5)
                   Family AEOLOTHRIPIDAE   Key II
-   Wings narrow, with at most 1 visible cross-vein (I.6)
                   Family THRIPIDAE   Key III

9   Compound eye with 3 or 4 facets (I.8) (these can be hard
    to count; each facet usually bulges out slightly in
    silhouette and has its own round area of pigment, but the
    areas of pigment often merge or are destroyed during
    specimen preparation); head without ocelli (simple eyes);
    wings absent   Larva   10
-   Compound eye with 5–60 or more facets (I.9); head
    usually with ocelli (simple eyes, fig. 33); wings absent or
    short   Adult   33

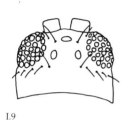

I.9

10 Antenna with segment V (counting from the head) less
    than half the length of segment IV   11
-   Antenna with segment V at least half the length of
    segment IV   28

11 Abdominal segments IV–VIII with 1 pair of setae on the
    underside; abdominal segment IX with 3 (female) or 4
    (male) pairs of setae (count setae on upperside and
    underside, note that one pair is much shorter than the
    others) (pl. 3.2)   Larva I, family THRIPIDAE
                                  (see Speyer & Parr, 1941)
-   Abdominal segments IV–VIII with 3 pairs of setae on the
    underside; abdominal segment IX with 5 (female) or 6
    (male) pairs of setae (pl. 3.3)
                                  Larva II, family THRIPIDAE   12

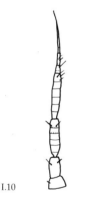

I.10

(The following couplets can distinguish some common species. For
further confirmation, check your specimen against the key and
description in Speyer & Parr (1941))

12 Antenna with tip segment more than 5 times as long as
    its width at the base of the segment (I.10) (check that you
    can count not more than 7 segments; faint lines across the
    segments can give the false impression of several short
    segments at the tip)
                   Sub-family PANCHAETOTHRIPINAE   13
-   Antenna with tip segment less than 5 times as long as its
    width at the base of the segment (I.11)
                   Sub-family THRIPINAE   15

I.11

13 Abdominal segments IX and X partly darkened on the
    upperside (this can be faint; it shows up best if white
    paper is placed beneath the slide)   14
-   Abdominal segments IX and X not darkened on the
    upperside   other species

14 Abdominal segment II with a pair of spiracles (breathing pores), about 20 µm in diameter, one on each side of the segment   *Hercinothrips femoralis* (Reuter)
– Abdominal segment II without a pair of spiracles
     Heliothrips haemorrhoidalis (Bouché)

15 Abdominal segment IX with a row of teeth along the upperside hind edge   16
– Abdominal segment IX without a row of teeth along the upperside hind edge   27

16 Abdominal segment IX at least partly darkened on the upperside (darkened areas usually appear light brown or grey in contrast with the surroundings, but can be faint; they show up best if white paper is placed beneath the slide)   17
– Abdominal segment IX not darkened on the upperside 25

17 Upperside without darkening on the head or thorax   18
– Upperside with some darkening on the head or thorax or both   20

18 Abdominal segment IX with more than the hind three quarters darkened on the upperside; live specimens under a hand lens appear yellow with a black tip to the abdomen   *Kakothrips pisivorus* (Westwood)
(This species is also known as *Kakothrips robustus* (Uzel))
– Abdominal segment IX with only the hind half or less darkened on the upperside   19

19 Abdominal segment IX with 8 teeth on the upperside hind edge, the longest being more than twice the length of the shortest   *Taeniothrips inconsequens* (Uzel)
– Abdominal segment IX with more than 10 teeth on the upperside hind edge, of similar size, the longest being less than twice the length of the shortest
     Frankliniella intonsa (Trybom)

20 Pronotum (upperside of segment behind head, fig. 33 p. 24) with at least a quarter of its area darkened; head with at least half the area of the upperside darkened   21
– Pronotum with less than a quarter of its area darkened; head with less than half the area of the upperside darkened   22

plaque

I.12

21 Abdominal segments VII and VIII with at least some
    setae more than 40 µm long on the upperside
                                   Thrips angusticeps Uzel
 – Abdominal segments VII and VIII with no setae more
    than 40 µm long on the upperside   *Thrips validus* Uzel

22 Abdominal segment II with a pair of spiracles
    (breathing pores), about 20 µm long, one on
    each side of the segment   23
 – Abdominal segment II without a pair of spiracles
                                   Thrips tabaci Lindeman

I.13

23 Abdominal segments V–VII with at least some plaques
    (I.12) on the upperside; pronotum with 1–8 pairs of
    darkened spots   24
 – Abdominal segments V–VII without plaques on the
    upperside; pronotum with 2–4 pairs of darkened spots
                                   Thrips physapus L.

I.14

24 Pronotum with 1–2 pairs of darkened spots
                                   Thrips vulgatissimus Haliday
 – Pronotum with 4–8 pairs of darkened spots
                                   Thrips atratus Haliday

25 Abdominal segments VII and VIII with some setae at
    least 40 µm long on the upperside   *Thrips flavus* Schrank
 – Abdominal segments VII and VIII with all setae
    less than 40 µm long on the upperside   26

I.15

26 Thoracic segments II and III with plaques (I.12) on the
    underside extended into extremely small spines (I.13) (to
    see this, you need a well-mounted specimen and a well-
    adjusted microscope at x1000 with oil immersion)
                                   Thrips major Uzel
 – Thoracic segments II and III with plaques on the
    underside not extended into extremely small spines (I.14)
                                   Thrips fuscipennis Haliday

I.16

27 Abdominal segment VII with setae 1 on the underside
    more than 35 µm long and longer than setae 2 and 3 (I.15)
                                   Limothrips cerealium Haliday
 – Not like this   other species

28 Antennal segments III and IV without rings or rows of
    very small spines; compound eyes with 3 facets   29
 – Antennal segments III and IV with at least 3 rings or with
    at least 1 row of very small spines (I.16); compound eyes
    with 4 facets   31

29 Antennal segment III less than 1.5 times as long as its maximum width (measure the entire length, including any narrow neck at the base); abdominal segment IX with 3 (female) or 4 (male) pairs of setae (count setae on upperside and underside)
Larva I, family PHLAEOTHRIPIDAE

– Antennal segment III more than 1.5 times as long as its maximum width; abdominal segment IX with 5 pairs of setae. Larva II, family PHLAEOTHRIPIDAE 30

30 Body colour red except for the rear half of abdominal segment IX, the whole of segment X and 2 large plates on the upperside of the segment behind the head, which are brown to black (the plates may be hard to see in specimens on microscope slides); legs and antennae brown to black; found in flowers of ox-eye daisy (*Leucanthemum vulgare*)
probably *Haplothrips leucanthemi* (Schrank)

– Not like this other species

31 Abdominal segments IV–VIII with 1 pair of setae on the underside; abdominal segment IX with 3 (female) or 4 (male) pairs of setae (count setae on upperside and underside) Larva I, family AEOLOTHRIPIDAE
(see Speyer & Parr, 1941)

– Abdominal segments IV–VIII with 3 pairs of setae on the underside; abdominal segment IX with 5 (female) or 6 (male) pairs of setae
Larva II, family AEOLOTHRIPIDAE 32

I.17

32 Abdominal segment IX with a row of 4 pegs of similar length on the upperside (I.17); pronotum (upperside of segment behind head) with 4 pairs of dark patches (these show up best if white paper is placed beneath the slide)
possibly *Aeolothrips intermedius* Bagnall
(Speyer & Parr (1941) give a detailed description which can be used to confirm identification)

– Not like this other species

33 Antenna with 7 or 8 segments (examine the tip closely for small segments) Male or female 34

– Antenna with 9 segments Male 35

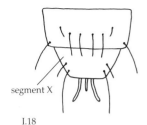

segment X

I.18

34   All setae on the upperside at the end of the abdomen arising from the tergite (upperside plate, fig. 33 p. 24) (I.18); abdominal segment X bluntly rounded (I.18)
           Adult male, family THRIPIDAE   Key III

–   Some setae on the upperside at the end of the abdomen arising beyond and separate from the hind edge of the tergite (I.19); abdominal segment X tubular and as if cut off at the tip (truncate) (I.19)   Adult male or female, family PHLAEOTHRIPIDAE   Key IV

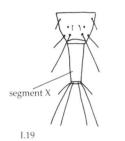

segment X

I.19

35   Antennal segments III and IV (counting from head) each with a branched sense organ (I.20) (these are hard to see because they are almost transparent; high magnification and a well-adjusted microscope are needed)
           Family THRIPIDAE   Key III

–   Antennal segments III and IV each with an unbranched sense organ (I.21, I.22)   36

36   Antennal segments III and IV with an elongate sense organ not projecting from the surface (I.21)
           Family AEOLOTHRIPIDAE   Key II

–   Antennal segments III and IV with an elongate sense organ projecting from the surface (I.22)
           Family THRIPIDAE   Key III

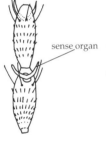

sense organ

I.20

37   Antennae projecting forwards or sideways (pl. 3.4)   38

–   Antennae projecting backwards over (I.23, pl. 4.1) or alongside (I.24) the head   39

38   Antennae less than 3 times as long as their width half-way along, projecting sideways; wing pads (developing wings) absent   Propupa, family PHLAEOTHRIPIDAE

–   Antennae more than 3 times as long as their width half-way along, projecting forwards or sideways; wing pads absent or short   Propupa, family THRIPIDAE

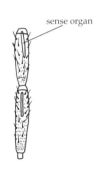

sense organ

I.21

39   Antennae projecting backwards along the side of the head (check that your specimen is not crushed by the coverslip, because this could push antennae from the top of the head to the side) (I.24)   40

–   Antennae projecting backwards over the top of the head (I.23, pl. 4.1)   41

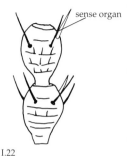

1.22

1.23

1.24

40 Antennae not reaching back beyond the front edge of the pronotum (upperside of segment behind head); wing pads (developing wings), if present, never reaching beyond abdominal segment II
     Pupa I, family PHLAEOTHRIPIDAE

–   Antennae reaching back beyond and underneath the front edge of the pronotum (I.24); wing pads, if present, sometimes reaching beyond abdominal segment II
     Pupa II, family PHLAEOTHRIPIDAE

41 Antennae reaching back beyond the hind edge of the pronotum (upperside of segment behind head)
     Pupa, family AEOLOTHRIPIDAE

–   Antennae not reaching back beyond the hind edge of the pronotum (pl. 4.1)   42

42 Surrounded by a loosely woven cocoon; wing pads (developing wings) absent or short (not reaching beyond abdominal segment III)
     Propupa, family AEOLOTHRIPIDAE

–   Not surrounded by a cocoon; wing pads absent, short or long (reaching beyond abdominal segment III)
     Pupa, family THRIPIDAE

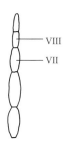

II.1

II Adults in the family Aeolothripidae

1 Antenna with indentations between the last 3 segments
   (segments VII–IX, counting from the head) (II.1)   2
– Antenna without indentations between the last 3
   segments (II.2, II.3)   3

2 Forewing with 2 grey-brown bands across its width
                                     Melanthrips ficalbii Buffa
– Forewing entirely grey-brown, apart from a pale
   area at the base   *Melanthrips fuscus* (Sulzer)

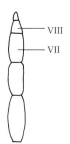

II.2

3 Antenna with indentations between segments V and VI,
   and VI and VII (II.2)   *Rhipidothrips* species
– Antenna without indentations between segments V and
   VI, and VI and VII (II.3)   4

4 Head, thorax and abdominal segments I–III (counting
   from the thorax) with at least some areas yellow or white
                                  other *Aeolothrips* species
– Head, thorax and abdominal segments I–III
   uniformly brown   5

5 Forewing with 1 dark band across its width
                             Aeolothrips vittatus Haliday
– Forewing with 2 dark bands across its width, the bands
   often merging towards the hind edge of the wing   6

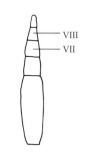

II.3

6 Forewing with the 2 dark bands merging towards the
   hind edge of the wing   Other *Aeolothrips* species
– Forewing with the 2 dark bands not merging   7

7 Wing-tip vein (II.4) darker than the surrounding
   membrane, about as dark as the veins in the dark bands
   (this shows up best if white paper is placed beneath
   the slide)   8
– Wing-tip vein as pale as the surrounding membrane   9

8 Antennal segments III and IV each with an elongate
   sense organ along the surface that is more than half as
   long as the segment (II.5)

   (these are hard to see because they are almost transparent; high
   magnification and a well-adjusted microscope are needed)
                             Aeolothrips propinquus Bagnall
– Antennal segments III and IV each with an elongate
   sense organ along the surface that is up to half as long as
   the segment   *Aeolothrips tenuicornis* Bagnall

II.4

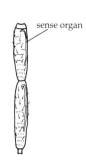

sense organ

9 Antennal segments I and most of II brown, darker than the base of segment III   *Aeolothrips intermedius* Bagnall

– Antennal segments I and II white-yellow, similar in colour to the base of segment III

Aeolothrips ericae Bagnall

II.5

III Adults in the family Thripidae

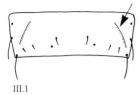

III.1

Tergites are plates on the upperside of the abdomen and sternites are plates on the underside of the abdomen (fig. 33 p. 24). They are numbered according to the segment they are on. To locate a particular numbered abdominal segment, it is usually easiest to find the last obvious visible segment at the end of the body, which is segment X (fig. 33 p. 24), and then count down from X, forwards along the body. For some species, it is useful to measure the lengths of setae in micrometres, so you may need an eyepiece graticule that has been calibrated against a stage micrometer.

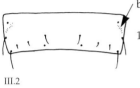

III.2

1   Abdominal tergite VII (plate on the upperside of abdominal segment VII) with one or more rows of at least 6 very small spines on each side (high magnification needed) (III.1, III.2, III.3)   2

–   Abdominal tergite VII without a row of at least 6 very small spines on each side   10

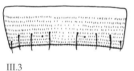

III.3

2   Abdominal tergite VII with 1–4 rows of very small spines on each side, covering a total of less than one third of the area of the tergite (III.1, III.2)   3

–   Abdominal tergite VII with at least 6 rows of very small spines on each side, covering a total of more than one third of the area of the tergite (III.3)   8

3   Antenna with 7 segments (examine the tip closely for small segments)   4

–   Antenna with 8 segments   6

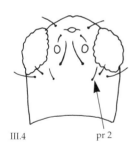

III.4   pr 2

4   Head with the second pair of setae from the centre in the row behind the compound eyes displaced backwards and out of line (labelled pr 2 in III.4)   5

–   Head with setae behind the compound eyes in a straight line (III.5)   *Thrips* species   50

5   Head length from level with the front of the compound eyes to the hind edge of the upper surface (measure along the mid-line of the body) longer than the maximum head width   *Baliothrips graminum* (Uzel)

(This species is also known as *Stenothrips graminum* Uzel)

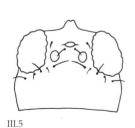

III.5

–   Head length from level with the front of the compound eyes to the hind edge of the upper surface shorter than the maximum head width   other *Baliothrips* species

(Some of these species are also considered to be *Stenothrips* species and *Stenchaetothrips* species)

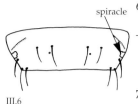

spiracle

III.6

6   Abdominal tergite V with at least one row of small
     spines on each side   7

–   Abdominal tergite V without a row of small spines
     on each side   *Kakothrips pisivorus* (Westwood)

     (This species is also known as *Kakothrips robustus* (Uzel))

III.7

7   Abdominal tergite VIII with rows of small spines in front
     of the spiracles (breathing pores) (III.6)
                                        Frankliniella species 46

–   Abdominal tergite VIII with rows of small spines
     between the spiracles (III.7)   *Thrips* species 50

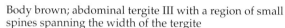

III.8

8   Body brown; abdominal tergite III with a region of small
     spines spanning the width of the tergite
                                            Sericothrips species

     (*Sericothrips abnormis* (Karny) lives on bird's-foot trefoil (*Lotus
     corniculatus*). *Sericothrips gracilicornis* Williams lives on tufted vetch (*Vicia
     cracca*). *Sericothrips staphylinus* Haliday lives on gorse (*Ulex* species))

–   Body yellow; abdominal tergite III with two regions of
     small spines, one on each side, not meeting in the
     middle of the tergite (III.8)   9

9   Antenna with 6 segments (examine the tip closely for
     small segments); found in the open
                                        Drepanothrips reuteri Uzel

–   Antenna with 8 segments; found in glasshouses
                                    Scirtothrips longipennis (Bagnall)

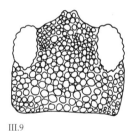

III.9

10   Upper surface of head covered with round pits (III.9);
     found in glasshouses   11

–   Upper surface of head not covered with round pits;
     found in the open or in glasshouses   12

11   Forewing with setae set in from the wing edge less than
     0.2 times as long as the width of the wing at the middle;
     at least one pair of legs not entirely yellow
                                Heliothrips haemorrhoidalis (Bouché)

–   Forewing with setae set in from the wing edge more than
     0.3 times as long as the width of the wing at the middle;
     all legs entirely yellow   *Helionothrips errans* (Williams),
                                    Parthenothrips dracaenae (Heeger)
                                        and *Hercinothrips* species

     (See Mound and others (1976) to identify these species further)

PLATE 1

**Some thrips and the damage
they cause**

1. *Frankliniella occidentalis*
   (adult female)

2. Front view of *Verbena*
   flower damaged by
   Frankliniella occidentalis

3. Side view of *Verbena* flower
   damaged by *Frankliniella
   occidentalis*

4. Tomato with tomato
   spotted wilt virus (TSWV)
   transmitted by *Frankliniella
   occidentalis*

5. *Thrips angusticeps*
   (adult female)

6. Lupin leaves damaged by
   Thrips angusticeps

7. Linseed stem damaged by
   Thrips angusticeps

8. *Kakothrips pisivorus*
   (larva II)

9. Pea pod damaged by
   Kakothrips pisivorus

1, 5 and 8 are x 20 natural size;
2 and 3 are x 3;
4, 6, 7 and 9 are x 0.8

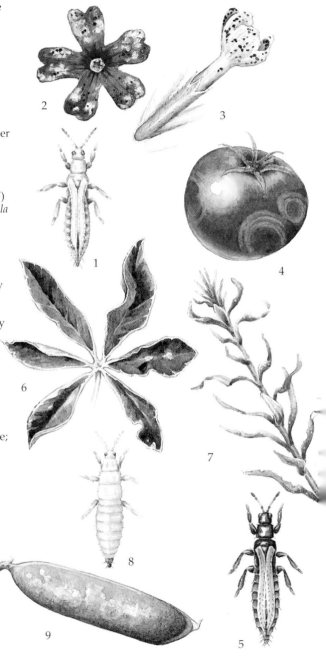

PLATE 2

**Haplothrips leucanthemi
on ox-eye daisy**

1. Adult feeding on pollen projecting from a floret

2. Adult feeding on pollen within a floret

3. Adult grooming wings

4. Male (on top) and female (below) mating

5. Adult female laying an egg in a floret

6. Florets containing eggs

7. Larva II

8. Adults on an ox-eye daisy

1–6 are x 16 natural size;
7 is x 19;
8 is natural size

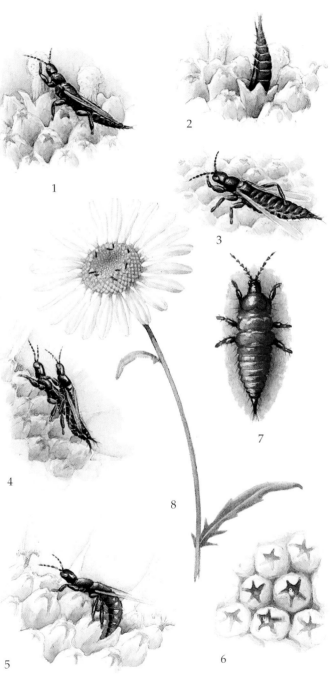

PLATE 3

Stages of *Frankliniella occidentalis*

1. Egg

2. Larva I

3. Larva II

4. Propupa

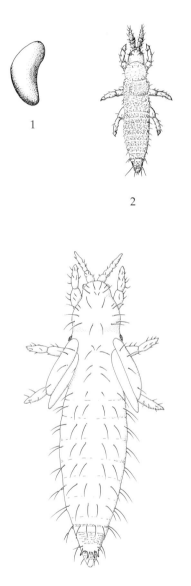
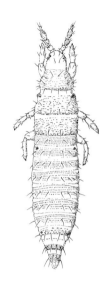

1

2

3

4

1 mm

PLATE 4

Stages of *Frankliniella occidentalis*

1. Pupa

2. Adult male

3. Adult female

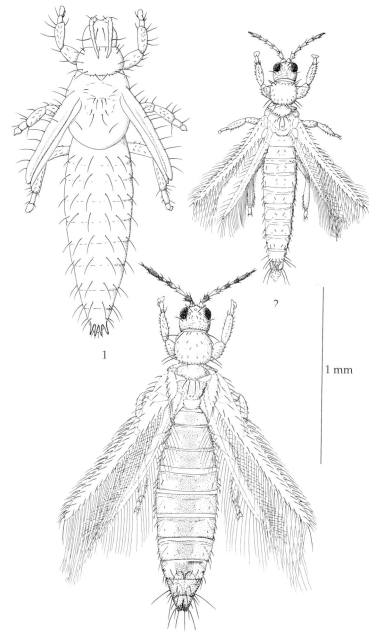

1 mm

1

2

3

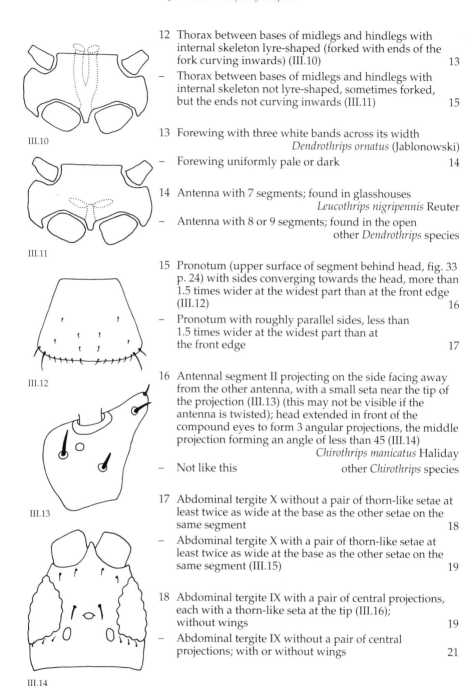

III.10

III.11

III.12

III.13

III.14

12 Thorax between bases of midlegs and hindlegs with
    internal skeleton lyre-shaped (forked with ends of the
    fork curving inwards) (III.10)   13
–   Thorax between bases of midlegs and hindlegs with
    internal skeleton not lyre-shaped, sometimes forked,
    but the ends not curving inwards (III.11)   15

13 Forewing with three white bands across its width
                              Dendrothrips ornatus (Jablonowski)
–   Forewing uniformly pale or dark   14

14 Antenna with 7 segments; found in glasshouses
                              Leucothrips nigripennis Reuter
–   Antenna with 8 or 9 segments; found in the open
                              other *Dendrothrips* species

15 Pronotum (upper surface of segment behind head, fig. 33
    p. 24) with sides converging towards the head, more than
    1.5 times wider at the widest part than at the front edge
    (III.12)   16
–   Pronotum with roughly parallel sides, less than
    1.5 times wider at the widest part than at
    the front edge   17

16 Antennal segment II projecting on the side facing away
    from the other antenna, with a small seta near the tip of
    the projection (III.13) (this may not be visible if the
    antenna is twisted); head extended in front of the
    compound eyes to form 3 angular projections, the middle
    projection forming an angle of less than 45 (III.14)
                              Chirothrips manicatus Haliday
–   Not like this   other *Chirothrips* species

17 Abdominal tergite X without a pair of thorn-like setae at
    least twice as wide at the base as the other setae on the
    same segment   18
–   Abdominal tergite X with a pair of thorn-like setae at
    least twice as wide at the base as the other setae on the
    same segment (III.15)   19

18 Abdominal tergite IX with a pair of central projections,
    each with a thorn-like seta at the tip (III.16);
    without wings   19
–   Abdominal tergite IX without a pair of central
    projections; with or without wings   21

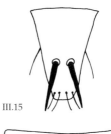

III.15

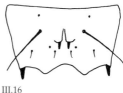

III.16

19 Antennal segment III projecting on the side facing away from the other antenna (III.17)
    Limothrips denticornis Haliday
– Antennal segment III not projecting on the side facing away from the other antenna (III.18, III.19) 20

20 Antennal segments III and IV each with a forked sense cone (III.18) *Limothrips schmutzi* Priesner
– Antennal segments III and IV each with an unforked sense cone (III.19) (fig. 15 p. 10)
    Limothrips cerealium Haliday

21 Antennal segments III and IV each with an unforked sense cone (III.19); without wings 22
– Antennal segments III and IV each with a forked sense cone (III.18); with or without wings 26

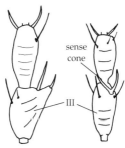

sense cone

III

III.17   III.18

22 Head short, length from level with the front of the compound eyes to the hind edge of the upper surface (measure along the mid-line of the body) less than the maximum width; body dark brown
    Apterothrips secticornis (Trybom)
– Head long, length from level with the front of the compound eyes to the hind edge of the upper surface more than the maximum width; body golden yellow or dark brown 23

23 Antenna with 6 segments (examine the tip closely for small segments) 24
– Antenna with 8 segments or, in rare specimens, 7 segments 25

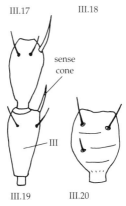

sense cone

III

III.19   III.20

24 Antennal segment II narrowing gradually on both sides towards base of segment (III.20)
    Aptinothrips rufus (Haliday)
– Antennal segment II narrowing abruptly on both sides (III.21) *Aptinothrips elegans* Priesner

25 Antennal segment II narrowing gradually on both sides towards base of segment (III.20); tarsi (fig. 33 p. 24) with 2 segments *Aptinothrips stylifer* Trybom
– Antennal segment II narrowing abruptly on one (III.22) or both (III.21) sides; tarsi with 1 segment
    other *Aptinothrips species*

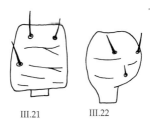

III.21   III.22

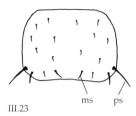

III.23   ms   ps

26 Pronotum with at least 1 pair of long setae at the hind corners (labelled ps in III.23) (long setae are either longer than half the length of the pronotum or, if shorter than this, at least 1.5 times as long as the pair of setae in the middle of the hind edge (labelled ms in III.23); if setae are close to the boundary between long and short, check that they are lying flat, otherwise they appear too short and give a misleading result, so identification will not be reliable)   27

–   Pronotum without long setae at the hind corners   42

27 Pronotum with 1 pair of long setae at the hind corners
          Dichromothrips orchidis Priesner,
          Tmetothrips subapterus (Haliday)
          and *Oxythrips* species
          (See Mound and others (1976) to identify these species further)

–   Pronotum with 2 or more pairs of long setae at the hind corners   28

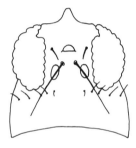

III.24

28 Pronotum with 2 pairs of long setae at the hind corners   29

–   Pronotum with 3 pairs of long setae at the hind corners
          Scolothrips longicornis Priesner

29 Antenna with 7 segments (examine the tip closely for small segments)   *Platythrips tunicatus* (Haliday)
          and *Bolacothrips jordani* Uzel

–   Antenna with 8 segments   30

30 Upperside of head with 1 or 2 pairs of setae in the region between the compound eyes and between or in front of the rear ocelli (fig. 33 p. 24, III.24) (focus carefully to ensure that you are not counting any setae on the underside of the head)   31

–   Upperside of head with 3 pairs of setae in the region between the compound eyes and between or in front of the rear ocelli (III.25)   34

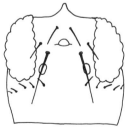

III.25

31 Upperside of head with 1 pair of setae in the region between the compound eyes and between or in front of the rear ocelli; antennal segments VII and VIII together longer than segment VI
          Rhaphidothrips longistylosus Uzel

–   Upperside of head with 2 pairs of setae in the region between the compound eyes and between or in front of the rear ocelli (III.24); antennal segments VII and VIII together shorter than segment VI   32

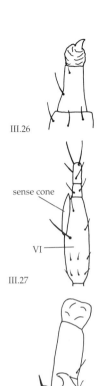

III.26

sense cone

VI

III.27

III.28

III.29

III.30

32 Body yellow; found in glasshouses
Chaetanaphothrips orchidii (Moulton)
– Body brown; found in the open 33

33 Front leg with a claw at the tip (III.26)
Taeniothrips inconsequens (Uzel)
– Front leg without a claw at the tip
Taeniothrips picipes (Zetterstedt)

34 Antennal segment VI with an enlarged sense cone lying
against the inner surface of the segment for most of its
length (III.27) (the inner surface is the side facing the
other antenna); front leg often with one or two claws at
the tip of the tibia (fig. 33 p. 24, III.28, III.29)
Odontothrips species 35

(Pitkin (1972) gives further details of identification of *Odontothrips*
species)

– Antennal segment VI without an enlarged sense cone;
front leg never with a claw at the tip of the tibia 40

35 Front leg with one or two claws at the tip of the tibia
(fig. 33 p. 24, III.28, III.29) 36
– Front leg without claws at the tip of the tibia
other *Odontothrips* species

36 Front leg with a curved claw and a non-curved
projection at the tip of the tibia (III.28) 37
– Front leg with two curved claws at the tip of the tibia
(III.29) (focus up and down carefully; one claw may
overlap the other) 38

37 Antennal segment IV intermediate in colour between
segments III and V; forewing vein 1 with 16–22 setae in
females and 14–20 setae in males (see couplet 59 to
separate males and females) *Odontothrips loti* (Haliday)
– Antennal segment IV the same colour as segment V and
darker than segment III; forewing vein 1 with 12–17 setae
in females and 10–15 setae in males
Odontothrips phaleratus (Haliday)

38 Abdominal tergites II to VIII each with lines in the area
immediately between the middle pair of setae; antennal
segment IV intermediate in colour between segments III
and V *Odontothrips biuncus* John
– Abdominal tergites II to VIII each without lines in the
area immediately between the middle pair of setae
(III.30); antennal segment IV the same colour as
segment V and darker than segment III 39

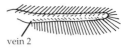

vein 2

III.31

39 Forewing vein 2 (III.31) with 16 to 25 setae; found on
    gorse (*Ulex*)   *Odontothrips ulicis* (Haliday)
–   Forewing vein 2 with 13 to 19 setae; found on broom
    (*Cytisus*)   *Odontothrips cytisi* (Haliday)
    (These species cannot always be separated. Pitkin (1972) gives details of
    how to identify the males from their reproductive organs)

40 Abdominal tergite VIII with a comb of setae along the
    whole length of the hind edge   *Mycterothrips* species
–   Abdominal tergite VIII without a comb of setae along the
    hind edge or with a comb of setae that has a gap in
    the middle   41

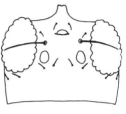

III.32

41 Antennal segment III at least as dark as segment II; head
    with the pair of setae immediately in front of the hind
    ocelli longer than the distance between the inner edges of
    the hind ocelli (III.32)   *Ceratothrips ericae* (Haliday)
–   Antennal segment III paler than segment II; head with
    the pair of setae immediately in front of the hind ocelli
    usually shorter than the distance between the
    inner edges of the hind ocelli   *Ceratothrips frici* (Uzel)

42 Antennal segments VII and VIII together more than 0.75
    times as long as segment VI (check that the antennae
    have 8 segments; if they seem to have 9 you may be
    mistaking a groove on segment VI for a gap between
    segments)   *Belothrips* species
–   Antennal segments VII and VIII together less than 0.5
    times as long as segment VI   43

III.33

43 Abdominal tergite VIII with a comb of spines or teeth
    on the hind edge (III.33, III.34)   44
–   Abdominal tergite VIII without a comb of spines or
    teeth on the hind edge   other *Anaphothrips* species

III.34

44 Abdominal tergite VIII with a comb of teeth on the
    hind edge (III.33)   *Anaphothrips articulosus* Priesner
–   Abdominal tergite VIII with a comb of spines on the
    hind edge (III.34)   45

45 Body mostly yellow with a dark band across the head;
    found in the open   *Anaphothrips obscurus* (Müller)
–   Not like this   other *Anaphothrips* species

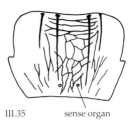

III.35   sense organ

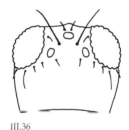

III.36

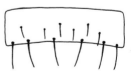

III.37

III.38

46 Metanotum (upperside of the third segment behind the head, fig. 33 p. 24) with the middle pair of setae less than half as long as the length of the metanotum
Frankliniella iridis (Watson)

–   Metanotum with the middle pair of setae more than two thirds as long as the length of the metanotum (III.35)   47

47 Metanotum with a pair of small, round sense organs (III.35); head with longest setae behind the compound eyes at least half as long as the setae within the triangle formed by the three ocelli   western flower thrips
Frankliniella occidentalis Pergande

–   Metanotum without a pair of small, round sense organs; head with longest setae behind the compound eyes usually less than half as long as the setae within the triangle formed by the three ocelli   48

48 Head with setae close together within the triangle formed by the three ocelli, the distance between the bases (not the sockets) of the setae less than a third of the length of the setae   *Frankliniella schultzei* (Trybom)

–   Head with setae far apart within the triangle formed by the three ocelli, the distance between the bases of the setae more than a third of the length of the setae (III.36)   49

49 Head long, the length from the front of the head between the antennae to the hind edge of the upper surface more than 0.8 times as long as the maximum width
Frankliniella tenuicornis (Uzel)

–   Head short, the length from the front of the head between the antennae to the hind edge of the upper surface less than 0.8 times as long as the maximum width
Frankliniella intonsa (Trybom)

50 Abdominal sternites III and IV (plates on the underside of abdominal segments III and IV, fig. 33 p. 24) with additional setae besides the 6 setae along the hind edge (III.37) (focus carefully to ensure you are not looking at the setae on the upperside)   51

–   Abdominal sternites III and IV without additional setae besides the 6 setae along the hind edge   62

51 Pleurotergites (plates on either side of the tergite, fig. 33 p. 24) on abdominal segments III and IV with additional setae besides the one on the hind edge (III.38)   52

–   Pleurotergites on abdominal segments III and IV without additional setae besides the one on the hind edge   56

52 Antenna with 7 segments (examine the tip closely for
small segments)   53
–   Antenna with 8 segments   54

53 Forewing vein 1 with 3 setae in the outer half (III.39)
Thrips pillichi Priesner
–   Forewing vein 1 with 7–11 setae in the outer half
Thrips minutissimus L.

vein 1

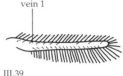

III.39

54 Forewing vein 1 with 3 setae in the outer half (III.39);
forewing pale   *Thrips vulgatissimus* Haliday
–   Forewing vein 1 with at least 5 setae in the outer half;
forewing mostly dark   55

55 Pronotum (upper surface of segment behind head, fig. 33
p. 24) with 1 or 2 pairs of setae on the front edge more
than 2.0 times as long as the setae in the central area of
the pronotum (check that you are not looking at setae on
the underside)   *Thrips verbasci* (Priesner)
–   Pronotum with no setae on the front edge more than 1.5
times as long as the setae in the central area of the
pronotum   *Thrips atratus* Haliday

56 Antenna with 7 segments (examine the tip closely for
small segments)   57
–   Antenna with 8 segments   *Thrips simplex* (Morison)
(This species is also known as *Taeniothrips simplex* (Morison))

57 Forewings short, less than 2.5 times as long as the
maximum width of the thorax
short-winged form of *Thrips angusticeps* Uzel
–   Forewings long, more than 2.5 times as long as the
maximum width of the thorax   58

58 Forewing vein 1 with 3 or 4 setae in the outer
half (III.39)   59
(The following 4 species can be difficult to separate. Males of *Thrips
origani* have not been found in Britain. Males of *Thrips calcaratus* have
not been described)
–   Forewing vein 1 with 5–11 setae in the outer half
Thrips angusticeps Uzel

III.40

59 With a Y-shaped ovipositor (egg-laying apparatus) in
segments VIII–X of the abdomen (III.40)   females 60
–   Without a Y-shaped ovipositor in segments VIII–X
of the abdomen   males 61

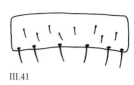

III.41

60 Abdominal sternites III–VI (plates on the underside of abdominal segments III–VI, fig. 33 p. 24) with at least 6 additional setae besides the 6 setae along the hind edge (III.41); abdominal tergite VIII with a comb of closely spaced spines on the hind edge, the distance between the spines in the middle of the comb always less than the length of the spines

Thrips hukkineni Priesner and *Thrips physapus* L.

(See Pitkin (1976) to separate females of these species. *Thrips hukkineni* is also known as *Thrips trehernei* Priesner)

– Not like this

Thrips calcaratus Uzel and *Thrips origani* Priesner

61 Body yellow   most probably *Thrips physapus* L., but possibly *Thrips origani* Priesner or *Thrips calcaratus* Uzel

– Body brown   most probably *Thrips hukkineni* Priesner, but possibly *Thrips calcaratus* Uzel

(*Thrips hukkineni* is also known as *Thrips trehernei* Priesner)

III.42

62 Pleurotergites (plates on either side of the tergite, fig. 33 p. 24) on abdominal segments III and IV with additional setae besides the one on the hind edge (focus carefully; do not confuse these setae with the seta at the edge of the tergite just next to the pleurotergite, see fig. 33 p. 24) (III.42)   other *Thrips* species

– Pleurotergites on abdominal segments III and IV without additional setae besides the one on the hind edge   63

III.43

63 Abdominal tergite VIII with a complete comb of spines or teeth along the whole length of the hind edge (III.43) (the comb is obvious when present in females, but is small and sparse in males, so use high magnification; see couplet 71 to separate the sexes)   64

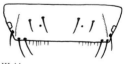

III.44

– Abdominal tergite VIII with a comb of spines or teeth along the hind edge that has a gap in the middle (III.44) or without a comb   71

64 Abdominal tergite VIII with a comb of spines (III.43) or a comb of triangular teeth tipped with spines   65

– Abdominal tergite VIII with a comb of triangular teeth not tipped with spines   71

III.45

65 Pleurotergites (plates on either side of the tergites, fig. 33 p. 24) covered with many rows of small spines (III.45)   66

– Pleurotergites without rows of small spines   67

vein 1

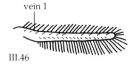

III.46

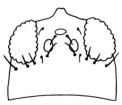

III.47

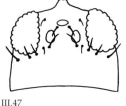

III.48

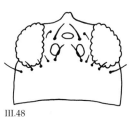

III.49

III.50

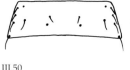

III.51

66   Forewing vein 1 usually with 4 setae in the outer half, sometimes with 3, 5 or 6 (III.46); antennal segments III–VI strongly 2-coloured, with yellow base and brown tip; rare, found on wood spurge (*Euphorbia amygdaloides*)   *Thrips euphorbiicola* Bagnall

–   Forewing vein 1 with 5–9 setae in the outer half; antennal segments III–VI uniformly coloured or gradually paler towards the base; common, found on many plants   *Thrips tabaci* Lindeman

67   Head and thorax yellow or almost clear, sometimes with brown markings   68

–   Head and thorax brown   70

68   Head with bases of a pair of setae lying within the triangle formed by the 3 ocelli (III.47); body yellow without brown or grey markings   *Thrips flavus* Schrank

–   Head without bases of a pair of setae lying within the triangle formed by the 3 ocelli (III.48); body yellow, often with brown or grey markings   69

69   Abdominal tergite V with the pair of setae nearest the middle more than 0.3 times as long as the length of the long setae at the hind corners (check that the setae are lying flat, otherwise they appear too short and give a misleading result); forewings usually short, not reaching beyond abdominal tergite II   *Thrips nigropilosus* Uzel

–   Abdominal tergite V with the pair of setae nearest the middle less than 0.3 times as long as the length of the long setae at the hind corners; forewings always long, reaching beyond abdominal tergite II   *Thrips alni* Uzel,   *Thrips urticae* F.   and *Thrips palmi* Karny

(*T. palmi* is not yet recorded from Britain, but it may arrive soon. It cannot be easily separated from the other two species. *T. alni* breeds on alder, *T. urticae* breeds on nettles, and *T. palmi* is a pest of many plants in glasshouses)

70   Metanotum (upperside of the third segment behind the head, fig. 33 p. 24) with lines behind the middle pair of setae enclosing areas nearly all more than twice as long as wide (III.49); abdominal tergite II with 4 setae along the edge of each side (III.50)   *Thrips validus* Uzel

–   Metanotum with lines behind the middle pair of setae enclosing areas mostly less than twice as long as wide (III.51); abdominal tergite II with 3 or 4 setae along the edge of each side   other *Thrips* species

(*Thrips linarius* will also key out at this couplet. It has a metanotum as described for *T. validus*, but abdominal tergite II has only 3 setae along the edge of each side. It has not been recorded from Britain, but it is a pest of flax and linseed on the Continent and might arrive here)

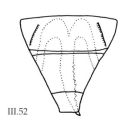

71 With a Y-shaped ovipositor (egg-laying apparatus) in
    segments VIII–X of the abdomen (III.52); females   72
–   Without a Y-shaped ovipositor in segments VIII–X
    of the abdomen; males   75

72 Body yellow; in flowers of hops (*Humulus lupulus*)
                                       Thrips albopilosus Uzel
III.52
–   Body brown or yellow and brown   73

73 Forewings short, not reaching beyond abdominal
    tergite II   other *Thrips* species
–   Forewings long, reaching beyond abdominal
    tergite II   74

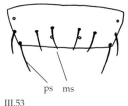

74 Abdominal tergite VII with setae at the hind corners
    (labelled ps in III.53) less than twice as long as the pair of
    setae nearest the middle of the tergite (labelled ms in
    III.53) (check that the setae are lying flat, otherwise they
    appear too short and give a misleading result)
ps ms   other *Thrips* species
III.53
–   Abdominal tergite VII with setae at the hind corners
    more than twice as long as the pair of setae nearest the
    middle of the tergite   76

75 Wings very short, not reaching beyond abdominal
    tergite II   other *Thrips* species
–   Wings long or half length, reaching beyond
    abdominal tergite II   76

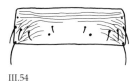

III.54

76 Abdominal tergites III–VII with no lines in the area
    immediately between the pair of setae nearest the middle
    of the tergite (III.54) (focus carefully, check that you are
    not looking at lines on the underside); pronotum (upper
    surface of segment behind head, fig. 33 p. 24) with
    continuous lines running across most of the width of
    the segment (III.55)   77
–   Not like this; found on willows (*Salix* species)
                                          other *Thrips* species

III.55

77 Abdominal tergite II with 3 setae along the edge of
    each side   78
–   Abdominal tergite II with 4 setae along the edge of
    each side (III.50)   79

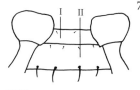

III.56

78 Abdominal sternite I with 2 or 3 very small (5 µm long) setae (III.56) (this sternite is small and these setae are very hard to see unless the specimen is well mounted); antennal segment III brown, the same colour as segment II (this shows up best at low magnification with white paper placed beneath the slide); rare, on juniper (*Juniperus communis*)   *Thrips juniperinus* L.

–   Abdominal sternite I without setae; antennal segment III yellow to brown, usually paler than segment II; very common, on flowers and leaves of many plants
                                                      Thrips major Uzel

79 Antennal segments III–V changing gradually from pale yellow to light brown, with segment V paler than segment VI (this shows up best at low magnification with white paper placed beneath the slide); uncommon, on flowers and leaves of elder (*Sambucus nigra*)
                                                      Thrips sambuci Uzel

–   Antennal segments III–V changing gradually from yellowish brown to dark brown, with segment V the same colour as segment VI; very common, on flowers and leaves of many plants   *Thrips fuscipennis* Haliday

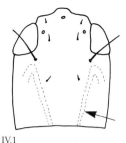

IV.1

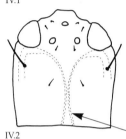

IV.2

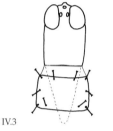

IV.3

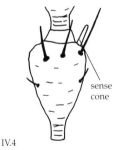

sense
cone

IV.4

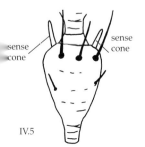

sense
cone

sense
cone

IV.5

IV Adults in the family Phlaeothripidae

Tergites are plates on the upperside of the abdomen and sternites are plates on the underside of the abdomen (fig. 33 p. 24). They are numbered according to the segment they are on. Note that abdominal tergite I is much smaller than the other tergites. To locate a particular numbered abdominal segment, it is easiest to find the tubular segment at the end of the body, which is segment X, and then count down from X, forwards along the body. You will need to measure lengths in micrometres on your specimen to use this key, so you will need an eyepiece graticule that has been calibrated against a stage micrometer. Lengths in millimetres are body lengths.

1 Head with maxillary stylets (retracted mouthparts) less than 5 μm wide (measure where indicated by arrow in IV.1 or IV.2; high magnification needed; focus carefully because a blurred image will appear too wide); 1.3–4.0 mm long (excluding antennae and bristles)
Sub-family PHLAEOTHRIPINAE 2

– Head with maxillary stylets more than 5 μm wide; 1.7–7.0 mm Sub-family IDOLOTHRIPINAE 34

2 Mouthcone on underside of head extending backwards beyond the hind edge of the pronotum (upperside of segment behind head) (IV.3); pronotum with white markings; 2.0–2.5 mm *Poecilothrips albopictus* Uzel

– Mouthcone not extending beyond the hind edge of the pronotum; pronotum without white markings; 1.3–4.0 mm 3

3 Wings absent or short, not reaching beyond abdominal tergite II (when counting tergites, note that the first abdominal tergite is much smaller than the others) 4

– Wings long, reaching beyond abdominal tergite II 6

4 Compound eyes reaching further back on the underside of the head than on the upperside (look at the outer surface and not the pigment underneath; compare the two surfaces by focusing up and down); antennal segment III with 1 sense cone, on the outer surface of the segment (IV.4) (high magnification needed; the outer surface is the side away from the other antenna); 1.4–2.1 mm *Cephalothrips monilicornis* (Reuter)

– Compound eyes not reaching further back on the underside of the head than on the upperside; antennal segment III with 2 or 3 sense cones, one on the inner surface and one or two on the outer surface of the segment (IV.5); 1.3–4.0 mm 5

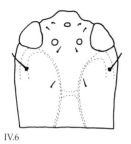

IV.6

IV.7

5   Head with maxillary stylets linked by a bridge (IV.6);
     living in flowers
                    Haplothrips statices (Haliday) form *morisoni*
–   Head with maxillary stylets not linked by a bridge;
     living on dead wood   *Hoplothrips* species   25

6   Forewings narrowed in the middle of the wing, with an
     additional row of 4–50 fringe hairs on the hind edge
     (IV.7) (these hairs stand out because they cross the other
     hairs on the forewing, but do not confuse them with
     hindwing hairs underneath)   7
–   Forewings with parallel sides near the middle of the
     wing, with or without an additional row of fringe hairs
     on the hind edge   18

7   Head with maxillary stylets linked by a bridge (IV.6).
     Haplothrips species   8
–   Head with maxillary stylets not linked by a bridge
                              Hoplandrothrips bidens (Bagnall)

8   Antennal segment III with 0, 1 or 3 sense cones
     (high magnification needed)   other *Haplothrips* species
–   Antennal segment III with 2 sense cones, one on the inner
     surface and one on the outer surface (IV.5) (the outer
     surface is the side away from the other antenna)   9

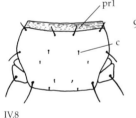

IV.8

9   Pronotum (upperside of segment behind head) with the
     pair of setae near the middle of the front edge (labelled
     pr 1 in IV.8) large, more than twice as long as the setae in
     the central area of the pronotum (labelled c in IV.8; check
     the setae are lying flat, otherwise they appear too short
     and identification will not be reliable)
                                    other *Haplothrips* species
     (*Haplothrips distinguendus* (Uzel) lives in flowers of *Carduus, Cirsium*
     and *Scrophularia. Haplothrips marrubiicola* Bagnall lives in flowers of
     Marrubium. Haplothrips senecionis Bagnall lives in flowers of *Senecio)*
–   Pronotum with the pair of setae near the middle of the
     front edge small, less than twice as long as the setae in
     the central area of the pronotum   10

10   Antennal segment IV with 2 sense cones, one inner
     and one outer   *Haplothrips fuliginosus* (Schille)
–   Antennal segment IV with 4 sense cones, two inner and
     two outer   11

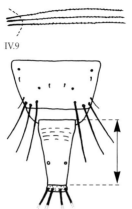

IV.9

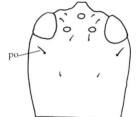

IV.10

11 Head with maxillary stylets close together, the shortest distance between them less than 0.2 times the maximum width of the head   12

– Head with maxillary stylets far apart, the shortest distance between them more than 0.2 times the maximum width of the head   13

12 Forewings colourless, except at the base; antennal segments IV–VI brown with yellow bases (this shows up best at low magnification with white paper beneath the slide; ignore the narrow neck at the base of the segments which is always clear); 1.6–2.6 mm
    Haplothrips juncorum Bagnall

– Forewings pale brown   13

13 Wing-tip fringe hairs with barbs (IV.9)
    Haplothrips setiger Priesner

– Wing-tip fringe hairs without barbs   14

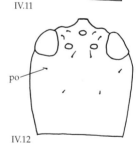

po

IV.11

14 Abdominal tergite X (the tube) less than 2.0 times as long as the width at the base (the specimen must not be crushed or the base will appear too wide; measure the maximum length parallel to the mid-line of the body, as in IV.10)   *Haplothrips hukkineni* Priesner

– Abdominal tergite X at least 2.4 times as long as the width at the base   15

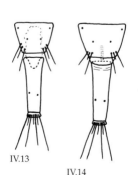

po

IV.12

15 Head with the pair of setae behind the compound eyes (labelled po in IV.11) longer than the width of the front ocellus (these setae are often hard to see in specimens that have not been macerated); found in flowers of thrift (*Armeria maritima*) or flowers of sheepsbit scabious (*Jasione montana*)   other *Haplothrips* species

– Head with the pair of setae behind the compound eyes not longer than the width of the front ocellus (IV.12)   16

16 Abdominal sternite X (plate on the underside of abdominal segment X) with front edge indented (IV.13);
    males *Haplothrips leucanthemi* (Schrank) and *Haplothrips propinquus* Bagnall

(Males of these two species are difficult to separate. See Mound and others (1976) for details)

– Abdominal sternite X with front edge not indented (IV.14); females   17

IV.13

IV.14

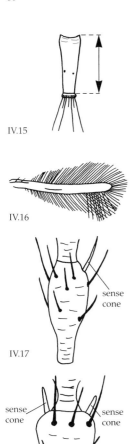

IV.15

IV.16

IV.17

sense
cone

sense
cone

sense
cone

IV.18

17 Abdominal tergite X usually 2.4–2.8 times as long as the width at the base (the specimen must not be crushed or the base will appear too wide; measure the maximum length parallel to the mid-line of the body, as in IV.15); found in flowers of yarrow *(Achillea millefolium)*
          Haplothrips propinquus Bagnall

– Abdominal tergite X usually 2.9–4.1 times as long as the width at the base; found in flowers of ox-eye daisy *(Leucanthemum vulgare)* *Haplothrips leucanthemi* (Schrank)

(Females of these two species are difficult to separate. If you wish to distinguish them with certainty, consult an expert)

18 Forewings with an additional row of 4–50 fringe hairs on the hind edge (IV.16)   19

– Forewings without an additional row of fringe hairs on the hind edge   24

19 Antennal segment III with 1 sense cone, on the outer surface (IV.17) (high magnification needed; the outer surface is the side away from the other antenna); living on tissues of live plants   20

– Antennal segment III with 2 or 3 sense cones, one on the inner surface and one or two on the outer surface of the segment (IV.18); living on dead wood   21

20 Antennal segment IV short, less than 1.8 times as long as its maximum width; living on lily bulbs in glasshouses; 2.0–2.5 mm   *Liothrips vaneeckei* Priesner

– Antennal segment IV long, more than 2.0 times as long as its maximum width; living on leaves of ash *(Fraxinus* species) or elm *(Ulmus* species); 2.3–3.7 mm
          Liothrips setinodis (Reuter)

21 Abdominal segments III–VIII with a white spot on each side; 3.0–4.0 mm (larva, fig. 10 p. 7)
          Acanthothrips nodicornis (Reuter)

– Abdominal segments III–VIII without white spots   22

22 Abdominal tergite IX with the longest pair of setae less than half as long as the length of tergite X
          Phlaeothrips species

– Abdominal tergite IX with the longest pair of setae more than half as long as the length of tergite X   23

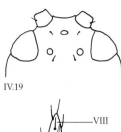

IV.19

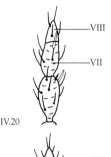

IV.20

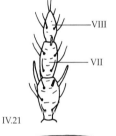

IV.21

IV.22

IV.23

IV.24

23 Antennal bases far apart, the distance between the base segments at their closest point at least as long as the maximum width of the segments (IV.19); antennal segments VII and VIII (the two segments at the tip) without a distinct indentation between them (IV.20)
Abiastothrips schaubergeri (Priesner)

(This species is also known as *Holothrips schaubergeri* (Priesner))

– Antennal bases close together, the distance between the base segments at their closest point shorter than the maximum width of the segments; antennal segments VII and VIII usually with a distinct indentation between them (IV.21)   25

24 Abdominal tergite IX with the longest pair of setae less than one third as long as the length of tergite X
Cephalothrips monilicornis (Reuter)

– Abdominal tergite IX with the longest pair of setae at least as long as the length of tergite X
Hoplothrips longisetis (Bagnall)

(This species is also known as *Maderothrips longisetis* (Bagnall))

25 Setae in the middle of the hind edge of tergite IX (labelled B1 in IV.22) at least 1.1 times as long as the length of tergite X   other *Hoplothrips* species

– Setae in the middle of the hind edge of tergite IX less than 0.9 times as long as the length of tergite X   26

26 Antennal segments IV and V each yellowish at the base, abruptly changing to dark brown in the half to two thirds nearer the tip (this shows up best at low magnification with white paper beneath the slide)   27

– Antennal segments IV and V each uniform in colour or gradually changing to brown towards the end nearer the tip   30

27 Abdominal sternite VIII (plate on the underside of abdominal segment VIII) with an irregularly shaped patch (a glandular area) across the width (IV.23, IV.24)   males 28

– Abdominal sternite VIII without an irregularly shaped patch across the width; females   29

28 Abdominal sternite VIII with the glandular area 15–20 μm wide (IV.23) (measured in the middle of the segment along the mid-line of the body)   *Hoplothrips ulmi* (F.)

– Abdominal sternite VIII with the glandular area 25–40 μm wide (IV.24)   *Hoplothrips fungi* (Zetterstedt)

29 Antennal segment III with the sense cone on the inner
surface up to 0.15 times as long as the setae in the middle
of the hind edge of abdominal tergite IX (measure the
sense cone along its length; high magnification needed;
the inner surface is the one facing the other antenna)
Hoplothrips ulmi (F.)
–   Antennal segment III with the sense cone on the inner
surface 0.16–0.25 times as long as the setae in the middle
of the hind edge of abdominal tergite IX
Hoplothrips fungi (Zetterstedt)

30 Antennal segments II and III equally pale yellow and
antennal segments IV and V equally brown (this shows
up best at low magnification with white paper beneath
the slide); antennal segment III with 2 sense cones (high
magnification needed)   *Hoplothrips semicaecus* (Uzel)
–   Not like this; antennal segment II often darker than
segment III; antennal segment III often the same colour
as segment IV; antennal segment III often with 3 sense
cones   31

31 Head yellow; pronotum (upperside of segment behind
head) brown (fig. 6 p. 5, fig. 7 p. 5)
Hoplothrips pedicularius (Haliday)
–   Head and pronotum the same colour, yellow or brown 32

IV.25   IV.26

32 Abdominal tergites VIII–X paler than abdominal tergites
I–VII; abdominal tergite VIII with the pair of setae at the
side blunt or asymmetrically pointed (IV.25) (high
magnification needed; this character is very difficult to
see); antennal segments II–V of similar colour or becoming
gradually darker towards the tip (fig. 6 p. 5, fig. 7 p. 5)
Hoplothrips pedicularius (Haliday)
–   Abdominal tergites VIII–X as dark as or darker than
abdominal tergites I–VII; abdominal tergite VIII with the
pair of setae at the side symmetrically pointed (IV.26);
antennal segment III often much paler than segments
IV and V; antennal segment II often darker than
segment III   33

33 Body length 1.3–2.0 mm; front leg with a tooth at the end
less than 0.3 times as long as the width of that part of the
leg (check that the tooth is mounted flat); rare, found on
dead wood from pine trees (*Pinus* species)
Hoplothrips polysticti (Morison)
–   Body length 2.5–4.0 mm; front leg with a tooth at the end
more than 0.5 times as long as the width of that part of
the leg; common, found on dead wood of many trees
Hoplothrips corticis (De Geer)

34 Head relatively short, length from level with the front of the compound eyes to the hind edge of the upper surface (measured along the mid-line of the body) less than 1.2 times as long as the maximum head width; 2.3–4.0 mm   35

– Head relatively long, length from level with the front of the compound eyes to the hind edge of the upper surface more than 1.3 times as long as the maximum head width; 1.7–7.0 mm   36

IV.27

35 Head with maxillary stylets close together, the shortest distance between them less than a quarter of the maximum width of the head; antennal bases far apart, the distance between the base segments at their closest point at least as long as the maximum width of the segments (IV.27); 2.3–3.8 mm   *Abiastothrips schaubergeri*

(This species is also known as *Holothrips schaubergeri* (Priesner))

– Head with maxillary stylets far apart, the shortest distance between them at least a quarter of the maximum width of the head; antennal bases close together, the distance between the base segments at their closest point shorter than the maximum width of the segments; 3.0–4.0 mm   *Bolothrips dentipes* (Reuter)

36 Body length less than 3.4 mm; head with maxillary stylets close together, the shortest distance between them less than a tenth of the maximum width of the head; 1.7–3.3 mm   *Cryptothrips nigripes* (Reuter)

– Body length more than 3.4 mm; head with maxillary stylets close together or far apart; 3.5–7.0 mm   37

37 Head with maxillary stylets occupying more than two thirds the length of the head; 3.5–4.5 mm   *Megalothrips bonannii* Uzel

– Head with maxillary stylets occupying less than two thirds the length of the head (IV.28); 3.5–7.0 mm   38

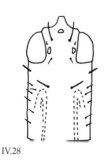

IV.28

38 Head relatively short, length from level with the front of the compound eyes to the hind margin of the upper surface less than 1.7 times as long as the maximum head width; 3.5–5.0 mm (fig. 12 p. 8)   *Megathrips lativentris* (Heeger)

– Head relatively long, length from level with the front of the compound eyes to the hind margin of the upper surface more than 1.7 times as long as the maximum head width; 4.0–7.0 mm   *Megathrips nobilis* Bagnall

(This species is also known as *Bacillothrips nobilis* (Bagnall))

7 Techniques

Finding thrips

Thrips are not difficult to find. However, a little experience is needed to know when and where to look for a particular species. The Royal Entomological Society Handbook on Thysanoptera (Mound and others, 1976) gives brief information about the plants on which each species of thrips can be found, the months in which male and female adults and larvae have been recorded and the counties where each species is known to occur. It can be used as a hunting guide. Thrips have been poorly collected in many areas of Britain, so the absence of a record from your county does not necessarily mean that a species is not present. Table 4 lists some common plants and the adult thrips that are commonest on them.

Thrips are perhaps easiest to find in flowers. Knocking a few flowers against the palm of your hand will often knock out a few thrips. The dedicated thrips collector always carries a beating tray, a stick, a fine paintbrush, a set of small tubes filled with 60% AGA (see below), and a hand lens. The stick is used to knock vegetation, sharply but not destructively, so that loose material falls onto the tray. After careful scrutiny of the debris, any thrips are transferred with the brush to a specimen tube. It is necessary to kill thrips to identify them, but if thrips are to be kept alive, they should be placed in a tight-sealing container, such as a plastic lunch box, with plenty of plant material, and kept at a constant temperature to avoid condensation. People who collect thrips spend a lot of time staring at a beating tray, waiting to see whether anything moves! A paper data label, written in pencil, with details of the date, place, plant species and plant part should be dropped into the tube with the specimens. The beating tray should be about 20–30 cm by 30–40 cm. A plastic photographic developing tray or a cut-down washing-up bowl or a thick cloth on a frame work well. Thrips show up well against a green or white background. Use of a beating tray allows the behaviour of the thrips to be observed briefly before they are transferred to specimen tubes, but the efficiency of extraction is not known so the method is not very good for quantifying thrips densities. When thrips are visible on flowers or leaves, they can be transferred on the end of a fine paintbrush directly to a specimen tube, without the need for a beating tray.

Grass or reeds can be swept with a sweep net. This catches many thrips and reveals which thrips are present. However, since the vegetation is usually a mixture of several species, it is not known which species of plant or which part of the plant the thrips were on before they were caught. Ecological information is lost.

Plant material can be collected directly into large specimen tubes half-filled with 60–70% alcohol and all the thrips removed later under a dissecting microscope. This method produces quantitative data, such as thrips densities per flower.

Thrips should preferably be collected in 60% AGA, which is 10 parts 60% alcohol (industrial methylated spirit diluted with water), 1 part glycerine and 1 part glacial acetic acid. The alcohol preserves the thrips, the glycerine softens the body and the acetic acid breaks down the body contents. As a substitute, 60–70% alcohol can be used; specimens collected in it do not stay as soft, but often this does not matter. Both substances are flammable and toxic. AGA should only be made up in a fume cupboard. In Britain, alcohol (industrial methylated spirit) may only be purchased with an Excise licence; local schools, universities or museums may have some. Surgical spirit (sold by chemists) and methylated spirits (sold by hardware shops) can be used if alcohol is not available, but they are far from ideal. They are about 90–95% alcohol, which will cause distortion of the specimen. If they are diluted with water, they precipitate a white substance which coats the specimen. Surgical spirit leaves an oily coating on specimens and methylated spirits contains a dye. It would be better to stick thrips onto card with a water-soluble glue and keep them dry than to store them in methylated spirits or surgical spirit.

Although specimens are best collected in AGA, they can be damaged if they are kept in it for more than a few weeks. If specimens are not going to be mounted promptly, they should be transferred to 60–70% alcohol, preferably after only a few hours in AGA, and kept in a cool dark place. The specimens should have their legs, wings and antennae spread out and their abdomen stretched at this stage, while they are still soft, so they can be mounted easily later on.

Fungus-feeding thrips are usually found on old wood or in leaf litter, because this is where the fungi are. They are hard to spot by eye. It is best to start by beating dead wood over a beating tray until some thrips are found and only then search by eye. Beating is more effective for twigs and small branches than for large logs. The thrips will be on or just under the bark. The wood should have been dead for some time, but should not be wet and crumbly. Leaf litter can be sorted over a beating tray, but thrips are extracted more efficiently with a Tullgren funnel. This is a large funnel with a coarse grid in it on which the leaf litter is placed. The leaf litter is warmed gently from above with a light bulb for about 12 hours so that thrips retreat down the funnel. A beaker of 60–70% alcohol is placed below the spout of the funnel to collect thrips that drop down.

Hoplothrips pedicularius is fairly common on dead wood of all kinds of deciduous trees. Try looking under broken branches below mature trees in oak woodland. The wood should not be too dry or too wet. The thrips live under *Stereum* fungus, which forms small brackets about 1–4 cm across. *Hoplothrips fungi* occurs in dense stands of young oak trees, where the lower branches have been killed by shading. It lives under the encrusting purplish grey *Peniophora* fungus.

It is easier, as well as better for the habitat, to search by beating rather than by peeling off fungi.

Many plant-feeding thrips are active from about April to September. Those that hibernate, such as *Limothrips cerealium*, can be extracted from leaf litter or loose bark near cereal fields throughout the winter and they will become active if brought into a warm room. Most fungus-feeding thrips can be found in every month of the year.

Observing thrips

Although thrips are not usually disturbed by being observed, they are often hidden from view behind leaves or petals. It may be necessary to create an artificial environment to be able to observe them more easily. Observations can be made indoors or outdoors.

It is particularly difficult to see how thrips feed because their mouthparts are usually obscured by the head and front legs. Pollen feeding can be observed by placing a thrips inside a clear gelatin capsule (size 0), as used for pills (from Farillon, address p. 62). A pin piercing the capsule at one end and held between thumb and forefinger is used to rotate the capsule so that the thrips is kept standing upside down on the upper surface and the mouthparts can be seen by looking through the transparent surface on which the thrips is standing (fig. 34). A good dissecting microscope and a cool light source are needed. If the inside of the capsule is smeared with pollen from a stamen, thrips can be observed feeding on pollen grains (Kirk, 1984b, 1987). Other types of transparent cell could be improvised, for example from a coverslip on a microscope cavity slide.

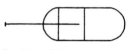

Fig. 34. A gelatin capsule used to observe thrips.

Thrips in open flowers can be observed feeding, fighting, mating and laying eggs. The flowers may even be picked and brought indoors without disturbing the thrips.

Fungus thrips can be observed feeding, fighting, mating and laying eggs on small pieces of fungi kept indoors in a small tight-sealing petri dish. A small piece of clear plastic can be sealed onto the fungus with clay to produce a small observation cell (Crespi, 1986a).

Trapping thrips

Large numbers of thrips can be caught in water traps and sticky traps. These are useful for experiments on host finding because the colour, scent and size of the trap can be varied. The disadvantages are that the hosts of the thrips are not known and the thrips are killed.

A water trap can be made from a plastic dish, such as a washing-up bowl or an empty ice-cream container. It is filled to about two centimetres below the rim with tap water containing a few drops of washing-up detergent. It is important to use an unscented detergent such as Teepol if you are investigating factors affecting trap catches. Thrips that land on the water sink to the bottom and drown. The catch must be sieved out within a few days or it will go mouldy. A piece of very fine cotton gauze can be glued across a hole in a plastic dish to form a sieve. The insects should be stored in 60–70% alcohol and sorted later under a dissecting microscope. Separating thrips from the other insects takes time. Watch out for thrips caught up in the bristles of flies and bees!

Sticky traps have the advantage that they can be left longer than water traps and they can be held vertically. They are useful when it is not necessary to identify the thrips, such as when only one species is likely to be caught, but they are a nuisance if thrips have to be mounted on microscope slides. The insects then have to be extracted from the sticky stuff and washed in a solvent before sorting and mounting. Sticky traps can be made by coating sheets of hard plastic with a sticky substance, such as Tanglefoot or Tangle-Trap (from the Tanglefoot Co., address p. 62), tree banding grease (from garden centres) or petroleum jelly (from chemists). Trap colours can be varied by painting traps with gloss paint.

Scent is released from small, glass tubes attached to the side of the trap and the release rate is controlled with a wick made from a cotton wool dental roll (from dental suppliers) (Kirk, 1985a). But beware, some of the scents that have been used to trap thrips are toxic and corrosive as liquids. They may need careful handling. Details of the hazards from individual chemicals are available from suppliers. Plant extracts can be made for scent experiments by crushing plant parts in methanol (methyl alcohol) as a solvent (from the Aldrich Chemical Co. Ltd, address p. 62). Methanol is flammable and highly toxic. Handle it in a fume cupboard or a well-ventilated space. Other solvents, such as industrial methylated spirits, could be tried, but the scent of some of the ingredients might affect the insects. Flowers or leaves could be tried in open tubes without a solvent, but the scent release rate would not be uniform.

Traps are often disturbed, or even shot at, if left unattended! It is best to keep them well out of the way of people and dogs. A chicken-wire cover may be necessary to keep off birds. Water traps are not very stable, so a roofing bolt is often pushed through a hole drilled in the middle of the base of the trap and poked into the ground or in the end of a bamboo cane to hold the trap in place. A nut and a thick rubber washer, which can be cut out of a rubber sheet, are screwed onto the bolt and tightened against the underside of the trap to form a waterproof seal between the bolt and the trap (fig. 35).

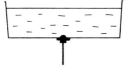

Fig. 35. A diagram of a cross-section of a water trap.

If you wish to investigate the effects of colour, scent and size, study papers describing such experiments for examples of experimental design and analysis (Kirk, 1984a, 1985a).

Mounting thrips

The full procedure for making high-quality permanent slides of thrips is as follows. Specimens can be manipulated under a dissecting microscope with fine watchmakers' forceps or a pair of mounted needles made by gluing a steel micropin to a matchstick. A pasteur pipette can be used to transfer liquids.

1. Transfer specimens in 60% AGA to a watch glass. Pipette off the liquid and replace it with 60% alcohol. Leave the specimens for at least 24 hours. The watch glass should be kept covered with a piece of glass to prevent losses from evaporation. This step is omitted if the specimens have been stored in 60% or 70% alcohol.
2. Replace the alcohol with 5% sodium hydroxide or potassium hydroxide. This breaks down the body contents of the specimens so that it is easier to see the surface structures. It also makes specimens paler. This is useful for very dark specimens on which surface structures do not show up well.
3. Pierce the body between the bases of the hind legs with a steel micropin to allow liquids to enter the body. Massage the abdomen slightly to expel some of the body contents.
4. After 30–240 minutes in the hydroxide, spread the wings and legs out and make sure the antennae are straight. Most British specimens will only need 30 minutes. Large, dark specimens should be left longer.
5. Replace the hydroxide with distilled water.
6. Add a little 60% alcohol to the water and then replace all the liquid with 60% alcohol.
7. Leave the specimens in 60% alcohol for at least 24 hours.
8. Replace the 60% alcohol with 70% alcohol and leave for 60 minutes. The alcohol concentration is being increased gradually to decrease the water content of the specimen. Keep fairly closely to the times given. The specimen may shrivel if left too long.
9. Replace the 70% alcohol with 80% alcohol and leave for 20 minutes.
10. Replace the 80% alcohol with 95% alcohol and leave for 10 minutes.
11. Replace the 95% alcohol with absolute alcohol and leave for 5 minutes.
12. Replace the absolute alcohol with fresh absolute alcohol and leave for another 5 minutes.
13. Replace the absolute alcohol with clove oil, terpineol or cedarwood oil. This acts as a clearing agent, an intermediary substance between the alcohol and the final mountant, and makes the specimen look transparent. After about 30 minutes, the specimens are ready to be mounted.
14. Place a small drop of canada balsam on a round coverslip (diameter 13 mm, thickness code 0). There should be enough balsam to spread to the edges of the coverslip when a microscope slide is placed on it. Place one thrips on its back in the drop on the coverslip. Spread the wings, legs and antennae.

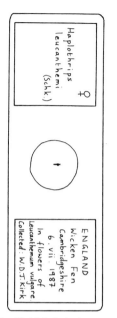

Fig. 36. A labelled microscope slide.

15. Lower a microscope slide onto the coverslip. As soon as the slide touches the balsam, turn it over. The coverslip should now be on top of the slide. This method tends to prevent air bubbles. If the mountant does not spread right to the edges of the coverslip, add more from the side.
16. Label the slide with the date and place of capture and details of where it was found (see fig. 36).
17. Place slides in an oven at about 37 °C or a warm place until the balsam is hard.

Slides prepared in this way will last a hundred years and would be welcomed by a museum. However, this laborious procedure may not always be possible or necessary. The use of sodium hydroxide can often be omitted for specimens that are not very dark because their surface structures will usually show up sufficiently well without it. Steps 1 to 7 would then be omitted, although the body should still be pierced as in step 3 to allow fast exchange of liquids.

Thrips can be mounted more quickly by transferring them from 60% AGA or 60–70% alcohol directly into a drop of an aqueous mounting medium on a coverslip. Follow steps 14 to 16, but use an aqueous mountant instead of canada balsam. Specimens may then have to be left for 2–3 days before the body surface becomes clearly visible. Possible mountants include polyvinyl lactophenol, Euparal, Hoyer's medium or Berlese fluid. These temporary slides last months to years before air bubbles appear and obscure the specimen.

Microscope slides, coverslips, mountants and other chemicals are available from suppliers (addresses p. 62). If the mountants become unavailable, it is usually possible to make them up from recipes in histology textbooks, such as Humason (1979). Mountants and their ingredients are often toxic and corrosive and need to be treated carefully. Ask for advice from suppliers.

Rearing thrips

It is difficult to rear individual thrips in small cages. They require high humidity, but will drown in small drops of condensation. If the cage is ventilated, they are liable to escape. The process is very labour-intensive. Lewis (1973) and Teulon (1992) give further details.

The effect of diet on egg laying can be investigated by keeping adult thrips in small cages for a few days. Small cages are easily made from polyethylene specimen tubes with caps (from BDH, address p. 62). A ventilation window about 10 mm square is cut in the side and covered with very fine cotton gauze, which is glued in place (fig. 37) (Kirk, 1985d). The bottom of the tube is cut off. Parafilm M (from BDH, address p. 62) is stretched as thin as possible (10–15 μm) and placed over the cut end. Clear adhesive tape is wrapped around, extending as a collar beyond this end of the tube, and

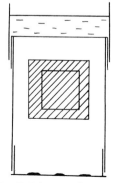

Fig. 37. A diagram of a cross-section of a small cage for thrips.

10% sucrose solution (10 g of granulated sugar made up to 100 ml of solution) is poured into the space formed, to a depth of about 5 mm. The lid of the tube is now the base of the cage. It can be opened to let in thrips and solid food, such as pollen. Thrips feed on the solution and lay eggs through the Parafilm membrane. Eggs laid in the sugar solution can be seen without opening the cage. The cages must be kept at a relative humidity of about 70–90%. One way to do this is to keep them in a sealed container above a layer of a saturated salt solution, such as potassium nitrate. Condensation can be avoided by keeping them at a constant temperature.

It is not difficult to rear large numbers of some pest species on plants kept indoors or in a glasshouse. For example, a few *Frankliniella occidentalis* will multiply rapidly when introduced to flowers of pot chrysanthemums. They may even infest other plants nearby! The culture can be kept going by placing fresh flowering plants next to the old plants as they finish flowering. The adult thrips will transfer themselves.

Writing papers

Writing up is an important part of a research project, particularly when the findings are to be communicated to other people. A really thorough, critical investigation that has established new information of general interest may be worth publishing if the insects and plants on which it is based can be identified with certainty. Journals that publish short papers on insect biology include the *Entomologist's Monthly Magazine*, the *Bulletin of the Amateur Entomologists' Society*, *The Entomologist* and, for projects with an educational slant, the *Journal of Biological Education*. Those unfamiliar with publishing conventions are advised to examine current numbers of these journals to see what sort of thing they publish, and then to write a paper along similar lines, keeping it short, but presenting enough information to establish the conclusions. It is then time to consult an appropriate expert who can give advice on whether and in what form the material might be published. It is an unbreakable convention of scientific publication that results are reported with scrupulous honesty. Hence it is essential to keep detailed and accurate records throughout the investigation, and to distinguish between certainty and probability, and between deduction and speculation. It will usually be necessary to apply statistical techniques to test the significance of the findings. A book such as Fowler & Cohen (1990) will help, but this is an area where expert advice can contribute much to the planning, as well as the analysis, of the work.

Table 4. *Some common plants and some of the adult thrips that are often found on them in Britain. The parts of the plant where the thrips can be found are abbreviated: L=leaves; F=flowers. An asterisk indicates that larvae are also likely to be found.*

Plant species	Common name	Part	Thrips species
Allium cepa	onion	L	*Thrips tabaci**
Brassica napus	oilseed rape	F	*Thrips vulgatissimus**
		F	*Aeolothrips intermedius*
Calluna vulgaris	heather	F	*Ceratothrips ericae**
Calystegia sepium	hedge bindweed	F	*Thrips fuscipennis*
		F	*Thrips major*
Chrysanthemum hybrids (in glasshouses)	pot chrysanthemum	F	*Frankliniella occidentalis**
Convolvulus arvensis	field bindweed	F	*Thrips atratus*
Cytisus scoparius	broom	F	*Odontothrips cytisi**
Erica cinerea	bell heather	F	*Ceratothrips ericae**
Leucanthemum vulgare	ox-eye daisy	F	*Haplothrips leucanthemi**
Ligustrum ovalifolium	garden privet	L	*Dendrothrips ornatus**
Linum usitatissimum	linseed	FL	*Thrips angusticeps**
Primula vulgaris	primrose	F	*Taeniothrips picipes**
Quercus robur	oak	F	*Thrips minutissimus**
Ranunculus acris	meadow buttercup	F	*Thrips flavus*
		F	*Thrips major*
		F	*Frankliniella intonsa*
Senecio jacobaea	ragwort	F	*Haplothrips senecionis**
Taraxacum officinale	dandelion	F	*Thrips physapus**
		F	*Thrips hukkineni**
		F	*Thrips tabaci**
		F	*Thrips validus**
		F	*Aeolothrips tenuicornis*
Triticum aestivum	wheat	FL	*Limothrips cerealium**
		FL	*Limothrips denticornis**
		FL	*Chirothrips manicatus**
		FL	*Baliothrips graminum**
Ulex europaeus	gorse	F	*Odontothrips ulicis**
		FL	*Sericothrips staphylinus**
		F	*Thrips flavus**
Urtica dioica	nettle	F	*Thrips urticae**
Vicia faba	field bean	F	*Kakothrips pisivorus**

Some useful addresses

Suppliers of equipment

Wide range of equipment, including polyethylene specimen tubes

BDH Laboratory Supplies, MERCK Ltd, Hunter Boulevard, Magna Park, Lutterworth, Leicestershire LE17 4XN

Flat-based specimen tubes for collecting and storing

Denley Instruments Ltd, Natts Lane, Billingshurst, West Sussex RH14 9EY

Empty clear gelatin capsules

Farillon Ltd, Ashton Road, Romford, Essex RM3 8UE

Dishes for water traps

A.W. Gregory & Co. Ltd, Glynde House, Glynde Street, London SE4 1RY

Microscopes and magnifiers

Hampshire Micro, The Microscope Shop, Oxford Road, Sutton Scotney, Hampshire SO21 3JG

Wide range of entomological equipment

Watkins and Doncaster, Conghurst Lane, Four Throws, Hawkhurst, Kent TN18 5ED

Suppliers of chemicals

Wide range of chemicals

Aldrich Chemical Co. Ltd, The Old Brickyard, New Road, Gillingham, Dorset SP8 4JL

BDH Laboratory Supplies, MERCK Ltd, Hunter Boulevard, Magna Park, Lutterworth, Leicestershire LE17 4XN

Microscopical chemicals

Northern Biological Supplies Ltd, 3 Betts Avenue, Martlesham Heath, Ipswich, Suffolk IP5 7HR

Timstar, Green Lane, Wardle, Nantwich, Cheshire CW5 6DB

Tanglefoot and Tangle-trap for sticky traps

Tanglefoot Company, 314 Straight Avenue SW, Grand Rapids, MI 49504, USA

Suppliers of new biological books

Naturalists' Handbooks and AIDGAP keys, including Tilling's key
Richmond Publishing Co. Ltd, P.O. Box 963, Slough SL2 3RS

Handbooks for the Identification of British Insects
The Royal Entomological Society, 41 Queen's Gate, London
SW7 5HU

Supplier of new and secondhand biological books

E.W. Classey Ltd, P.O. Box 93, Faringdon, Oxfordshire
SN7 7DR

Societies

Some readers may wish to join a society in order to
find out more about thrips and ecology and meet people
with similar interests. Most counties have a natural history
society. The names and addresses of some established
societies are given below.

The Amateur Entomologists' Society produces *A
Directory for Entomologists*, which contains a comprehensive
list of names and addresses of local and national
entomological societies and recording schemes, along with
many other useful addresses. It can be purchased from AES
Publications, The Hawthorns, Frating Road, Great Bromley,
Colchester, Essex CO7 7JN.

Amateur Entomologists' Society, 5 Oakfield, Plaistow,
West Sussex RH14 0QD

British Ecological Society, 26 Blades Court, Deodar Road,
Putney, London SW15 2NU

British Entomological and Natural History Society, c/o The
Royal Entomological Society, 41 Queen's Gate, London
SW7 5HU

Royal Entomological Society, 41 Queen's Gate, London
SW7 5HU

References and further reading

Finding books

There are very few books devoted solely to thrips. Lewis (1973) has been the standard work for a long time, but it is now out of print. Ananthakrishnan (1984a) deals mainly with Indian thrips. Two major new books are due out in the next few years. *Thrips as Crop Pests*, edited by T. Lewis, is to be published by CAB International in the UK, and an identification key to the thrips of Europe (in German) by R. zur Strassen will be published by Goecke & Evers in Germany.

Many of the books listed here will be unavailable in local and school libraries. It is possible to make arrangements to see or borrow them by seeking permission to visit the library of a local university, or by asking your local public library to borrow the work (or a photocopy of it) for you from the British Library Document Supply Centre. This may take several weeks, and it is important to present your librarian with a reference which is correct in every detail. References are acceptable in the form and order given here, namely the author's name and date of publication, followed by (for a book) the title and publisher or (for a journal article) the title of the article, the journal title, the volume number, and the first and last pages of the article.

The Handbooks for the Identification of British Insects are published by the Royal Entomological Society and can be bought by post from The Royal Entomological Society, 41 Queen's Gate, London SW7 5HU. Reprints of the key by Tilling are available by post from The Richmond Publishing Co. Ltd, P.O. Box 963, Slough SL2 3RS. Asterisks mark publications available from The Richmond Publishing Co. Ltd.

Ananthakrishnan, T.N. (1984a) *Bioecology of Thrips*. Oak Park, Michigan: Indira Publishing House.
Ananthakrishnan, T.N. (1984b) Biology of gall thrips (Thysanoptera: Insecta). In *Biology of Gall Insects*, ed. T.N. Ananthakrishnan, pp. 107–127. New Delhi: Oxford & IBH.
Andrewartha, H.G. & Birch, L.C. (1954) *The Distribution and Abundance of Animals*. Chicago: University of Chicago Press.
Begon, M., Harper, J.L. & Townsend, C.R. (1990) *Ecology: Individuals, Populations and Communities* (2nd edn). Boston, Massachusetts: Blackwell Scientific Publications.
Blum, M.S., Foottit, R. & Fales, H.M. (1992) Defensive chemistry and function of the exudate of the thrips *Haplothrips leucanthemi*. *Comparative Biochemistry and Physiology* **C102**, 209–211.
Chinery, M. (1976) *A Field Guide to the Insects of Britain and Europe* (2nd edn). London: Collins.
Chinery, M. (1986) *Collins Guide to the Insects of Britain and Western Europe*. London: Collins.
Chisholm, I.F. & Lewis, T. (1984) A new look at thrips (Thysanoptera) mouthparts, their action and effects of feeding on plant tissue. *Bulletin of Entomological Research* **74**, 663–675.
Crespi, B.J. (1986a) Territoriality and fighting in a colonial thrips, *Hoplothrips pedicularius*, and sexual dimorphism in Thysanoptera. *Ecological Entomology* **11**, 119–130.
Crespi, B.J. (1986b) Size assessment and alternative fighting tactics in *Elaphrothrips tuberculatus* (Insecta: Thysanoptera). *Animal Behaviour* **34**, 1324–1335.
Crespi, B.J. (1988) Risks and benefits of lethal male fighting in the colonial, polygynous thrips *Hoplothrips karnyi* (Insecta: Thysanoptera). *Behavioral Ecology and Sociobiology* **22**, 293–301.

Crespi, B.J. (1990) Subsociality and female reproductive success in a mycophagous thrips: an observational and experimental analysis. *Journal of Insect Behavior* **3**, 61–74.

Crespi, B.J. (1992a) Behavioural ecology of Australian gall thrips (Insecta, Thysanoptera). *Journal of Natural History* **26**, 769–809.

Crespi, B.J. (1992b) Eusociality in Australian gall thrips. *Nature* **359**, 724–726.

Cuthbertson, D.R. (1989) *Limothrips cerealium* – an alarming insect. *The Entomologist* **108**, 246–256.

Davidson, J. & Andrewartha, H.G. (1948) The influence of rainfall, evaporation and atmospheric temperature on fluctuations in the size of a natural population of *Thrips imaginis* (Thysanoptera). *Journal of Animal Ecology* **17**, 200–222.

Ellington, C.P. (1980) Wing mechanics and take-off preparation of *Thrips* (Thysanoptera). *Journal of Experimental Biology* **85**, 129–136.

Fowler, J. & Cohen, L. (1990) *Practical Statistics for Field Biology*. Milton Keynes: Open University Press.

Helyer, N.L. & Brobyn, P.J. (1992) Chemical control of western flower thrips (*Frankliniella occidentalis* Pergande). *Annals of Applied Biology* **121**, 219–231.

Heming, B.S. (1991) Order Thysanoptera. In *Immature Insects*. Vol. 2, ed. F.W. Stehr, pp. 1–21. Dubuque, Iowa: Kendall/Hunt Publishing Company.

Heming, B.S. (1993) Structure, function, ontogeny, and evolution of feeding in thrips (Thysanoptera). In *Functional Morphology of Insect Feeding*. Thomas Say Publications in Entomology, Proceedings. Eds. C.W. Schaefer & R.A.B. Leschen, pp. 3–41. Lanham, Maryland: Entomological Society of America.

Humason, G.L. (1979) *Animal Tissue Techniques* (4th edn). San Francisco: Freeman.

Kiester, A.R. & Strates, E. (1984) Social behaviour in thrips from Panama. *Journal of Natural History* **18**, 303–314.

Kirk, W.D.J. (1984a) Ecologically selective coloured traps. *Ecological Entomology* **9**, 35–41.

Kirk, W.D.J. (1984b) Pollen-feeding in thrips (Insecta: Thysanoptera). *Journal of Zoology, London* **204**, 107–117.

Kirk, W.D.J. (1985a) Effect of some floral scents on host finding by thrips (Insecta: Thysanoptera). *Journal of Chemical Ecology* **11**, 35–43.

Kirk, W.D.J. (1985b) Aggregation and mating of thrips in flowers of *Calystegia sepium*. *Ecological Entomology* **10**, 433-440.

Kirk, W.D.J. (1985c) Floral display in *Vicia faba*, and the distribution of a flower thrips, *Kakothrips pisivorus*. *Entomologia Experimentalis et Applicata* **38**, 233–238.

Kirk, W.D.J. (1985d) Pollen-feeding and the host specificity and fecundity of flower thrips (Thysanoptera). *Ecological Entomology* **10**, 281-289.

Kirk, W.D.J. (1987) How much pollen can thrips destroy? *Ecological Entomology* **12**, 31–40.

Lewis, T. (1959) The annual cycle of *Limothrips cerealium* Haliday (Thysanoptera) and its distribution in a wheat field. *Entomologia Experimentalis et Applicata* **2**, 187–203.

Lewis, T. (1964) The weather and mass flights of Thysanoptera. *Annals of Applied Biology* **53**, 165–170.

Lewis, T. (1973) *Thrips: their Biology, Ecology and Economic Importance*. London: Academic Press.

Loan, C. & Holdaway, F.G. (1955) Biology of the red clover thrips, *Haplothrips niger* (Osborn) (Thysanoptera: Phloeothripidae). *The Canadian Entomologist* **87**, 210–219.

Matteson, N.A. & Terry, L.I. (1992) Response to color by male and female *Frankliniella occidentalis* during swarming and non-swarming behavior. *Entomologia Experimentalis et Applicata* **63**, 187–201.

Mound, L.A. (1991) Secondary sexual character variation in male *Actinothrips* species (Insecta: Thysanoptera), and its probable significance in fighting behaviour. *Journal of Natural History* **25**, 933–943.

Mound, L.A., Morison, G.D., Pitkin, B.R. & Palmer, J.M. (1976) *Handbooks for the Identification of British Insects. Thysanoptera.* Vol. I, part 11. Royal Entomological Society of London.

Mound, L.A. & Palmer, J.M. (1983) The generic and tribal classification of spore-feeding Thysanoptera (Phlaeothripidae: Idolothripinae). *Bulletin of the British Museum (Natural History)* **46**(1), 1–174.

Mound, L.A. & Teulon, D.A.J. (1995) Thysanoptera as phytophagous opportunists. In *Thrips Biology and Management*, eds. B.L. Parker, M. Skinner & T. Lewis, pp. 3–19. New York: Plenum Press.

Parker, B.L., Skinner, M. & Lewis, T. (eds.) (1995) *Thrips Biology and Management*. New York: Plenum Press.

Pelikan, J. (1990) Butting in phlaeothripid larvae (Thysanoptera). *Proceedings of the 3rd International Symposium on Thysanoptera, Kazimierz Dolny, Poland, 1990*, 51–55.

Pelikan, J. (1994) Osmeterial glands in *Megathrips*-like males (Thysanoptera: Phlaeothripidae). *Courier Forschungsinstitut Senckenberg* **178**, 97–100.

*Redfern, M. & Askew, R.R. (1992) *Plant Galls*. Naturalists' Handbooks 17. Slough: The Richmond Publishing Co. Ltd.

Speyer, E.R. & Parr, W.J. (1941) The external structure of some Thysanopterous larvae. *Transactions of the Royal Entomological Society of London* **91**, 559–635.

Teerling, C.R., Pierce, H.D., Borden, J.H. & Gillespie, D.R. (1993) Identification and bioactivity of alarm pheromone in the western flower thrips, *Frankliniella occidentalis*. *Journal of Chemical Ecology* **19**, 681–697.

Terry, L.I. & Gardner, D. (1990) Male mating swarms in *Frankliniella occidentalis* (Pergande) (Thysanoptera: Thripidae). *Journal of Insect Behavior* **3**, 133–141.

Teulon, D.A.J. (1992) Laboratory techniques for rearing western flower thrips (Thysanoptera: Thripidae). *Journal of Economic Entomology* **85**, 895–899.

*Tilling, S.M. (1987) A key to the major groups of British terrestrial invertebrates. *Field Studies* **6**, 695–766.

*Unwin, D.M. & Corbet, S.A. (1991) *Insects, Plants and Microclimate*. Naturalists' Handbooks 15. Slough: The Richmond Publishing Co. Ltd.

Williams, C.B. (1915) The pea thrips (*Kakothrips robustus*). *Annals of Applied Biology* **1**, 227–246.

Wilson, E.O. (1971) *The Insect Societies*. Cambridge, Massachusetts: Belknap Press of Harvard University Press.

Index